The Gas Dynamics of Explosions

Explosions, and the non-steady shock propagation associated with them, continue to interest researchers working in different fields of physics and engineering (e.g., astrophysics, fusion, etc.). Based on the author's course in shock dynamics, this book describes the various analytical methods developed to determine non-steady shock propagation. These methods offer a simple alternative to the direct numerical integration of the Euler equations and offer a better insight into the physics of the problem. Professor Lee presents the subject systematically and in a style that is accessible to graduate students and researchers working in shock dynamics, combustion, high-speed aerodynamics, propulsion, and related topics.

Professor John H. S. Lee is Professor of Mechanical Engineering at McGill University, Montréal. He has been carrying out fundamental and applied research in combustion, detonations and shock wave physics, and explosion dynamics for the past forty years. As a consultant, Lee has also served on numerous government and industrial advisory committees on explosion hazards and safety. He has received many prizes, including the silver medal from the Combustion Institute (1980), the Dionizy Smoleński Medal from the Polish Academy of Sciences (1988), and the Nuna Manson gold medal (1991) for his outstanding contributions to the fundamentals and applied aspects of explosion and detonation phenomena. He is a fellow of Royal Society and the National Academy of Engineering of Canada.

The Gas Dynamics of Explosions

JOHN H. S. LEE

McGill University, Montréal, Canada

CAMBRIDGE
UNIVERSITY PRESS

CAMBRIDGE
UNIVERSITY PRESS

32 Avenue of the Americas, New York NY 10013

Cambridge University Press is part of the University of Cambridge.

It furthers the University's mission by disseminating knowledge in the pursuit of education, learning and research at the highest international levels of excellence.

www.cambridge.org
Information on this title: www.cambridge.org/9781107106307

First published 2016

A catalog record for this publication is available from the British Library

ISBN 978-1-107-10630-7 Hardback

Contents

Preface

Gas dynamics of explosions is a subject that deals with the non-steady propagation of shock waves. The theoretical description of non-steady shock propagation requires, in general, the solution of the non-linear partial differential equations of compressible flow that govern the unsteady flow behind the shock. This requires the numerical integration of the conservation equations. However, there exist various analytical methods that can give approximate solutions and provide a useful alternative to the more involved numerical integration of the gas dynamics equations. Analytical methods can also render the physics of the problem more transparent.

The material in this book is largely based on a course on shock dynamics that the author gave periodically since the 1970s. The objective of that course is to discuss the fundamentals of the non-steady gas dynamics and shock waves, where relatively few books on the subject are available. The choice of topics is that of the author and emphasis is placed on presenting the basics of gas dynamics. Thus, relatively few practical problems and numerical results are given, and sample problems are only used to serve as illustrations of the method.

Although the works of numerous authors are reviewed and developed upon, the book neither gives a comprehensive review of the extensive literature on the subject nor provides a detailed bibliography. It is felt that references can readily be obtained via an internet search. The few references that are given are mostly limited to the few early studies where the method was developed.

The author is extremely grateful to Dr. K. Ramamurthi who made valuable contributions and carried a very thorough reading of the manuscript. Dr. P. Thibault and Mr. M. Gaug also carried out proof reading of the manuscript. Drs. Qin Hui and Li Jian carried out the difficult task of typing the manuscript from the author's handwritten draft.

1 Basic Equations

1.1 Introduction

This chapter provides a brief review of unsteady gas dynamics and properties of shock and detonation waves. The medium of propagation of shocks and detonations is considered to be a perfect gas. The thermodynamics of perfect gases are first reviewed. The conservation equations in various variables (i.e., p, ρ, u, c, etc.) and in different coordinates systems, for example, Eulerian, Characteristic, Lagrangian, etc., are thereafter discussed. A brief discussion of waves of small amplitude (acoustic) and finite amplitude is given. Wave propagation is the foundation of non-steady compressible flows. The conservation equations across normal shock waves and the Rankine–Hugoniot relationships are then presented. This chapter is meant to provide the background to facilitate the reading of subsequent chapters.

1.2 Thermodynamics

We shall consider a perfect gas throughout this book. The equation of state for an ideal gas is given by

$$\begin{aligned} pv &= RT, \\ \text{or} \qquad p &= \rho RT, \end{aligned} \tag{1.2.1}$$

where $v = \frac{1}{\rho}$ is the specific volume, and $R = \frac{R_u}{M}$ is the gas constant, $R_u = 8.314\,\text{J/mol}\cdot\text{K}$ is the universal gas constant and M is the molecular weight. The relationship between the energy functions (internal energy e and enthalpy $h = e + pv$) and the state variables is known as the caloric equation of state. For an ideal gas, e and h are functions only of the temperature, and we write

$$\begin{aligned} e &= c_v T, \\ h &= c_p T, \end{aligned} \tag{1.2.2}$$

where c_v and c_p are functions of the temperature only. The relationship between c_p and c_v is

$$c_p - c_v = R. \tag{1.2.3}$$

The ratio of the specific heats is denoted by γ, that is,

$$\gamma = \frac{c_p}{c_v}. \tag{1.2.4}$$

In terms of p and ρ, Eq 1.2.2 can be written as

$$\begin{aligned} e = c_v T &= \frac{RT}{\gamma - 1} = \frac{pv}{\gamma - 1} = \frac{p}{\rho(\gamma - 1)}, \\ h = c_p T &= \frac{\gamma RT}{\gamma - 1} = \frac{\gamma pv}{\gamma - 1} = \frac{\gamma p}{\rho(\gamma - 1)}. \end{aligned} \tag{1.2.5}$$

The speed of sound is defined as

$$c^2 = \left(\frac{\partial p}{\partial \rho}\right)_s,$$

and for a perfect gas for which γ is constant, we write

$$c^2 = \gamma RT = \frac{\gamma p}{\rho}. \tag{1.2.6}$$

The entropy function, s, can be obtained from the relationship

$$T\mathrm{d}s = \mathrm{d}e + p\mathrm{d}v = \mathrm{d}h - v\mathrm{d}p. \tag{1.2.7}$$

For a perfect gas with constant γ, the above equations give,

$$\frac{s_2 - s_1}{R} = \ln\left(\frac{T_2}{T_1}\right)^{\frac{1}{\gamma-1}}\left(\frac{v_2}{v_1}\right) = \ln\left(\frac{T_2}{T_1}\right)^{\frac{\gamma}{\gamma-1}}\left(\frac{p_2}{p_1}\right). \tag{1.2.8}$$

For an isentropic process where $\mathrm{d}s = 0$, Eq 1.2.8 yields

$$\begin{aligned} T_2 v_2^{\gamma-1} &= T_1 v_1^{\gamma-1} = \frac{T}{\rho^{\gamma-1}} = \text{constant}. \\ \frac{T_2}{p_2^{\frac{\gamma-1}{\gamma}}} &= \frac{T_1}{p_1^{\frac{\gamma-1}{\gamma}}} = \frac{T}{p^{\frac{\gamma-1}{\gamma}}} = \text{constant}. \end{aligned} \tag{1.2.9}$$

Using the equation of state to eliminate T, we write

$$pv^\gamma = \frac{p}{\rho^\gamma} = \text{constant}. \tag{1.2.10}$$

The speed of sound can often be used to represent the energy function e and h. From Eqs 1.2.5 and 1.2.6, we obtain

$$e = \frac{c^2}{\gamma(\gamma - 1)} \qquad h = \frac{c^2}{\gamma - 1}. \tag{1.2.11}$$

1.3 Conservation Equations

We shall consider an inviscid, non-heat-conducting perfect gas. The conservation equations (Euler's equations) are given as

$$\frac{\mathrm{D}\rho}{\mathrm{D}t} + \rho\vec{\nabla}\cdot\vec{u} = 0, \tag{1.3.1}$$

$$\frac{\mathrm{D}\vec{u}}{\mathrm{D}t} + \frac{1}{\rho}\vec{\nabla}p = 0, \tag{1.3.2}$$

$$T\frac{\mathrm{D}s}{\mathrm{D}t} = \frac{\mathrm{D}e}{\mathrm{D}t} + p\frac{\mathrm{D}\frac{1}{\rho}}{\mathrm{D}t} = \frac{\mathrm{D}h}{\mathrm{D}t} - \frac{1}{\rho}\frac{\mathrm{D}p}{\mathrm{D}t} = 0, \tag{1.3.3}$$

where

$$\frac{\mathrm{D}}{\mathrm{D}t} = \frac{\partial}{\partial t} + \vec{u}\cdot\vec{\nabla}.$$

Equation 1.3.3 determines the change in entropy of a fluid particle as we follow its motion. When every particle has the same entropy, then $s =$ constant throughout. This is generally referred to as "homentropic" flow.

Using Eq 1.3.1 to eliminate $\frac{\mathrm{D}\rho}{\mathrm{D}t}$ in Eq 1.3.3, we obtain

$$\frac{\mathrm{D}e}{\mathrm{D}t} + \frac{p}{\rho}\vec{\nabla}\cdot\vec{u} = 0. \tag{1.3.4}$$

Expressing the internal energy in terms of the sound speed (Eq 1.2.11), Eq 1.3.4 can be written as

$$\frac{\mathrm{D}c}{\mathrm{D}t} + \left(\frac{\gamma-1}{2}\right)c\vec{\nabla}\cdot\vec{u} = 0. \tag{1.3.5}$$

The above equation is valid for particle isentropic flow. If this condition is not met, Eqs 1.3.3 and 1.3.5 give

$$\frac{\mathrm{D}c}{\mathrm{D}t} + \left(\frac{\gamma-1}{2}\right)c\vec{\nabla}\cdot\vec{u} = \left(\frac{\gamma-1}{2}\right)\frac{c}{R}\frac{\mathrm{D}s}{\mathrm{D}t}, \tag{1.3.6}$$

which is valid for non-isentropic flow. Using Eq 1.3.6, the momentum equation (Eq 1.3.4) can be written as

$$\frac{\mathrm{D}\vec{u}}{\mathrm{D}t} + \left(\frac{2c}{\gamma-1}\right)\vec{\nabla}c = T\vec{\nabla}s = \frac{c^2}{\gamma R}\vec{\nabla}s. \tag{1.3.7}$$

For particle isentropic flow or when there is no entropy gradient, where $\vec{\nabla}s = 0$, the right hand side of Eqs 1.3.6 and 1.3.7 vanishes.

Most of the time, we shall be considering one-dimensional non-steady flow with planar, cylindrical, and spherical symmetries. Equations 1.3.1 and 1.3.2 then become

$$\frac{\partial \rho}{\partial t} + u\frac{\partial \rho}{\partial r} + \rho\frac{\partial u}{\partial r} + \frac{j\rho u}{r} = 0, \tag{1.3.8}$$

$$\frac{\partial u}{\partial t} + u\frac{\partial u}{\partial r} + \frac{1}{\rho}\frac{\partial p}{\partial r} = 0, \tag{1.3.9}$$

where $j = 0, 1, 2$ for planar, cylindrical, and spherical geometries, respectively. Similarly, we may write Eqs 1.3.6 and 1.3.7 as

$$\frac{\partial c}{\partial t} + u\frac{\partial c}{\partial r} + \left(\frac{\gamma - 1}{2}\right)c\frac{\partial u}{\partial r} + \left(\frac{\gamma - 1}{2}\right)\frac{jcu}{r} = \left(\frac{\gamma - 1}{2}\right)\frac{c}{R}\frac{\mathrm{D}s}{\mathrm{D}t}, \tag{1.3.10}$$

$$\frac{\partial u}{\partial t} + u\frac{\partial u}{\partial r} + \frac{2c}{\gamma - 1}\frac{\partial c}{\partial r} = \frac{c^2}{\gamma R}\frac{\partial s}{\partial r}. \tag{1.3.11}$$

For particle isentropic flow, the right hand side of Eq 1.3.10 vanishes. If the entropy also is uniform, that is, $\frac{\partial s}{\partial r} = 0$, then the right hand side of Eq 1.3.11 also vanishes. Thus for homentropic flow, we write

$$\frac{\partial c}{\partial t} + u\frac{\partial c}{\partial r} + \left(\frac{\gamma - 1}{2}\right)c\frac{\partial u}{\partial r} + \left(\frac{\gamma - 1}{2}\right)\frac{jcu}{r} = 0, \tag{1.3.12}$$

$$\frac{\partial u}{\partial t} + u\frac{\partial u}{\partial r} + \frac{2c}{\gamma - 1}\frac{\partial c}{\partial r} = 0. \tag{1.3.13}$$

Equations 1.3.12 and 1.3.13 give a pair of equations for the dependent variables u and c. With c known, the pressure and density can be obtained from the isentropic relationships.

Equations derived previously (Eqs 1.3.8–1.3.13) are based on the Eulerian description where the flow variables are specified as a function of the position and time, that is, $p(r, t)$, $\rho(r, t)$, etc. An alternate description is the Lagrangian description where the fluid state of a particle is specified as a function of time. The Lagrangian variable, a, is generally chosen to be the initial position of the fluid particle. At a later time, t, the position is given by $r(a, t)$. The conservation equations for the general case of two- or three-dimensional flows in Lagrangian coordinates are quite complex. We shall consider only one-dimensional flow here. For simplicity, we shall derive the equations for spherical symmetry first and then generalize them for other geometries afterwards.

Let the initial position of a particular fluid particle at $t = 0$ be a. At a later time t, the position of the particle will be $r(a, t)$. Thus the mass in the spherical shell, $\rho_0 4\pi a^2 da$, at $t = 0$ will be $\rho 4\pi r^2 dr$ at a later time t. The conservation of mass is then given by

$$\rho_0 4\pi a^2 \mathrm{d}a = \rho 4\pi r^2 \mathrm{d}r(a, t). \tag{1.3.14}$$

Since

$$\mathrm{d}r(a, t) = \frac{\partial r}{\partial t}\mathrm{d}t + \frac{\partial r}{\partial a}\mathrm{d}a,$$

at constant time t, $\frac{dr}{da} = (\frac{\partial r}{\partial a})_t$, and the conservation of mass can be written as

$$\left(\frac{\partial r}{\partial a}\right)_t = \frac{\rho_0}{\rho}\left(\frac{a}{r}\right)^2. \tag{1.3.15}$$

The conservation equation for momentum of the particle at time t, can be written as

$$(\rho 4\pi r^2 dr)\left(\frac{\partial u}{\partial t}\right)_a = -4\pi r^2 dp,$$

and using Eq 1.3.15, the above expression becomes

$$\left(\frac{\partial u}{\partial t}\right)_a = -\frac{1}{\rho_0}\left(\frac{r}{a}\right)^2\left(\frac{\partial p}{\partial a}\right)_t. \tag{1.3.16}$$

For particle isentropic flow, the entropy of a fluid particle remains constant with time. Thus

$$\left(\frac{\partial s}{\partial t}\right)_a = 0. \tag{1.3.17}$$

Generalizing to planar and cylindrical geometries, we write Eqs 1.3.15 and 1.3.16 as

$$\frac{\partial r}{\partial a} = \frac{\rho_0}{\rho}\left(\frac{a}{r}\right)^j, \tag{1.3.18}$$

and

$$\frac{\partial u}{\partial t} = -\frac{1}{\rho_0}\left(\frac{a}{r}\right)^j\frac{\partial p}{\partial a}, \tag{1.3.19}$$

where $j = 0, 1, 2$ for planar, cylindrical, and spherical symmetries, respectively. It is understood that a is kept constant for the partial differentiation with respect to t and vice versa in Eqs 1.3.15.

1.4 Characteristic Equations

Euler equations are hyperbolic and have real characteristics. It is often convenient to integrate these equations along the characteristics. To obtain the conservation equations in characteristic form, we use the method of multipliers. Multiplying Eq 1.3.10 by α and Eq 1.3.11 by β and adding the resulting equations, we obtain after some re-arrangements, the following:

$$\left[\frac{\partial}{\partial t} + \left(u + \frac{\beta}{\alpha}\left(\frac{2}{\gamma-1}\right)c\right)\frac{\partial}{\partial r}\right]c + \frac{\beta}{\alpha}\left[\frac{\partial}{\partial t} + \left(u + \frac{\alpha}{\beta}\left(\frac{\gamma-1}{2}\right)c\right)\frac{\partial}{\partial r}\right]u$$

$$= -\frac{\gamma-1}{2}\frac{jcu}{r} + \frac{\gamma-1}{2}\frac{c}{R}\frac{\mathrm{D}S}{\mathrm{D}t} + \frac{\beta}{\alpha}\frac{c^2}{\gamma R}\frac{\partial s}{\partial r}. \tag{1.4.1}$$

We wish to write the above equation as a total differential equation of the form

$$\frac{\mathrm{d}c}{\mathrm{d}t}+\frac{\beta}{\alpha}\frac{\mathrm{d}u}{\mathrm{d}t}=-\frac{\gamma-1}{2}\frac{jcu}{r}+\frac{\gamma-1}{2}\frac{c}{R}\frac{\mathrm{D}s}{\mathrm{D}t}+\frac{\beta}{\alpha}\frac{c^2}{\gamma R}\frac{\partial s}{\partial r}. \tag{1.4.2}$$

Since c and u are functions of r and t, we write

$$\frac{\mathrm{d}c}{\mathrm{d}t}=\frac{\partial c}{\partial t}+\frac{\partial c}{\partial r}\left(\frac{\mathrm{d}r}{\mathrm{d}t}\right),$$

$$\frac{\mathrm{d}u}{\mathrm{d}t}=\frac{\partial u}{\partial t}+\frac{\partial u}{\partial r}\left(\frac{\mathrm{d}r}{\mathrm{d}t}\right).$$

From Eq 1.4.1, we see that

$$\frac{\mathrm{d}r}{\mathrm{d}t}=u+\frac{\beta}{\alpha}\left(\frac{2}{\gamma-1}\right)c=u+\frac{\alpha}{\beta}\left(\frac{\gamma-1}{2}\right)c.$$

Solving for the multipliers, we get

$$\frac{\beta}{\alpha}=\pm\frac{\gamma-1}{2},$$

and Eq 1.4.1 can then be written as

$$\left[\frac{\partial}{\partial t}+(u\pm c)\frac{\partial}{\partial r}\right]c\pm\frac{\gamma-1}{2}\left[\frac{\partial}{\partial t}+(u\pm c)\frac{\partial}{\partial r}\right]u$$

$$=-\frac{\gamma-1}{2}\frac{jcu}{r}+\frac{\gamma-1}{2}\frac{c}{R}\frac{\mathrm{D}s}{\mathrm{D}t}\pm\frac{\gamma-1}{2}\frac{c^2}{\gamma R}\frac{\partial s}{\partial r}. \tag{1.4.3}$$

The characteristic curves are given by

$$\frac{\mathrm{d}r}{\mathrm{d}t}=u\pm c. \tag{1.4.4}$$

We may write Eq 1.4.3 alternately as

$$\frac{\delta}{\delta t}\left(\frac{2c}{\gamma-1}\pm u\right)=-\frac{jcu}{r}+\frac{c}{R}\frac{\mathrm{D}s}{\mathrm{D}t}\pm\frac{\gamma-1}{2}\frac{c^2}{\gamma R}\frac{\partial s}{\partial r}. \tag{1.4.5}$$

where

$$\frac{\delta}{\delta t}=\frac{\partial}{\partial t}+(u\pm c)\frac{\partial}{\partial r}.$$

Defining $J^{\pm}=\frac{2}{\gamma-1}c\pm u$, Eq 1.4.5 may be written as

$$\frac{\delta}{\delta t}J^{\pm}=-\frac{jcu}{r}+\frac{c}{R}\frac{\mathrm{D}s}{\mathrm{D}t}\pm\frac{c^2}{\gamma R}\frac{\partial s}{\partial r}. \tag{1.4.6}$$

For particle isentropic flow $\left(\frac{\mathrm{D}s}{\mathrm{D}t}=0\right)$ and if the entropy is also uniform, then $\frac{\partial s}{\partial r}=0$ and Eq 1.4.6 reduces to

$$\frac{\delta J^{\pm}}{\delta t}=-\frac{jcu}{r}. \tag{1.4.7}$$

For planar geometry where $j = 0$, Eq 1.4.7 can be integrated to yield

$$J^{\pm} = \text{constant} = \frac{2}{\gamma - 1}c \pm u.$$

If $c = c_0$ when $u = 0$, then

$$J^{\pm} = \frac{2}{\gamma - 1}c \pm u = \frac{2}{\gamma - 1}c_0,$$

and solving for c gives

$$\begin{aligned} c &= c_0 \mp \frac{\gamma - 1}{2}u, \\ \text{or} \quad u &= \pm\frac{2}{\gamma - 1}(c_0 - c). \end{aligned} \tag{1.4.8}$$

The above equations give $c(u)$ or $u(c)$, and we may also express c and u in terms of the Riemann invariants $J^{\pm}$, that is,

$$\begin{aligned} u &= \left(\frac{J^+ - J^-}{2}\right), \\ c &= \frac{\gamma - 1}{2}\left(\frac{J^+ + J^-}{2}\right). \end{aligned} \tag{1.4.9}$$

Using the above equations we can write the characteristic equations (Eq 1.4.4) as

$$\frac{\mathrm{d}r}{\mathrm{d}t} = u + c = \left(\frac{J^+ - J^-}{2}\right) + \frac{\gamma - 1}{2}\left(\frac{J^+ + J^-}{2}\right), \tag{1.4.10}$$

$$\frac{\mathrm{d}r}{\mathrm{d}t} = u - c = \left(\frac{J^+ - J^-}{2}\right) - \frac{\gamma - 1}{2}\left(\frac{J^+ + J^-}{2}\right). \tag{1.4.11}$$

Since $J^+ =$ constant along a C^+, $\left(\frac{\mathrm{d}r}{\mathrm{d}t}\right)^+$ will depend on the value of J^-. Similarly for a C^-, the slope $\left(\frac{\mathrm{d}r}{\mathrm{d}t}\right)^-$ will depend on the value of the J^+. For a uniform flow where J^+ and J^- are constants, $\left(\frac{\mathrm{d}r}{\mathrm{d}t}\right)^{\pm} =$ constant and the characteristics become two sets of parallel lines.

For a non-steady flow adjacent to a uniform or stationary region, the Riemann invariant from the uniform region will be transmitted to the non-steady region via the characteristic. Thus one of the two Riemann invariants will be constant throughout both regions. Figure 1.1 shows the uniform and non-uniform regions. A J^- from the uniform region will be carried by a C^- characteristic into the non-steady region.

Thus $\frac{2}{\gamma-1}c - u$ is constant throughout and if $u = 0$, $c = c_0$ in the uniform region, that is, if

$$\frac{2}{\gamma - 1}c - u = \frac{2}{\gamma - 1}c_0,$$

$$\text{then} \quad c = c_0 + \frac{\gamma - 1}{2}u. \tag{1.4.12}$$

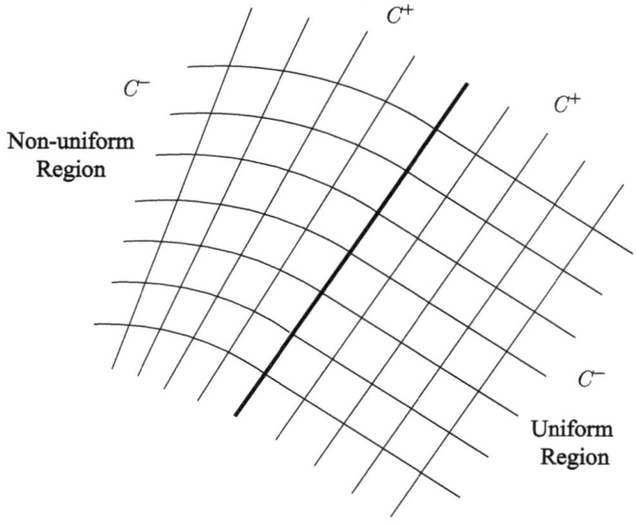

Figure 1.1 Uniform and Non-uniform Regions

Hence, a C^+ in the non-steady region will then be given by

$$\frac{dr}{dt} = u + c = c_0 + \frac{\gamma+1}{2}u,$$

which integrates to yield the trajectory of the C^+ as

$$r = \left(c_0 + \frac{\gamma+1}{2}u\right)t + f(u), \tag{1.4.13}$$

where the constant of integration $f(u)$ can be determined from initial or boundary conditions. Along a C^+, $J^+ = \frac{2}{\gamma-1}c + u =$ constant and using Eq 1.3.12, $J^+ = \frac{2}{\gamma-1}(2c - c_0)$. Thus, each C^+ is characterized by a value of u (or alternately a value of c).

Similarly, if a C^+ carries a J^+ from a uniform region into the non-uniform region and the uniform region is defined by $u = 0$, $c = c_0$, then

$$J^+ = \frac{2}{\gamma-1}c + u = \frac{2}{\gamma-1}c_0 = \text{constant},$$

and thus

$$c = c_0 - \frac{\gamma-1}{2}u. \tag{1.4.14}$$

A C^- in the non-uniform region is thus

$$\frac{dr}{dt} = u - c = -c_0 + \frac{\gamma+1}{2}u, \tag{1.4.15}$$

which integrates to yield

$$r = \left(-c_0 + \frac{\gamma + 1}{2} u \right) t + f(u). \tag{1.4.16}$$

Combining Eqs 1.4.13 and 1.4.16, we write

$$r = \left(\pm c_0 + \frac{\gamma + 1}{2} u \right) t + f(u), \tag{1.4.17}$$

for the trajectories of the $C^{\pm}$ in the non-uniform region adjacent to a uniform region where $u = 0$, $c = c_0$.

For the particular case where the characteristics originate from a point (e.g., $x = 0$ at $t = 0$), then $f(u) = 0$ and Eq 1.4.17 becomes

$$r = \left(\pm c_0 + \frac{\gamma + 1}{2} u \right) t. \tag{1.4.18}$$

Solving for u, we get

$$\frac{u}{c_0} = \frac{2}{\gamma + 1} \left(\frac{x}{c_0 t} \mp 1 \right). \tag{1.4.19}$$

From Eqs 1.4.12 and 1.4.14, we get

$$\begin{aligned} \frac{c}{c_0} = 1 \pm \frac{\gamma + 1}{2} \frac{u}{c_0} &= 1 \pm \left(\frac{\gamma - 1}{\gamma + 1} \right) \left(\frac{x}{c_0 t} \mp 1 \right) \\ &= \frac{2}{\gamma + 1} \left(1 \pm \frac{\gamma - 1}{2} \frac{x}{c_0 t} \right). \end{aligned} \tag{1.4.20}$$

The pressure, density, and temperature can be obtained using the isentropic relationships and Eq 1.4.20, that is,

$$\frac{p}{p_0} = \left(\frac{T}{T_0} \right)^{\frac{\gamma}{\gamma - 1}} = \left(\frac{c}{c_0} \right)^{\frac{2\gamma}{\gamma - 1}},$$

$$\frac{\rho}{\rho_0} = \left(\frac{T}{T_0} \right)^{\frac{1}{\gamma - 1}}.$$

Equations 1.4.12, 1.4.14, and 1.4.17 describe what is referred to as a "simple wave" flow.

1.5 Acoustic Waves

Small flow perturbations propagate as acoustic waves governed by the linear wave equation. Writing $p = p_0 + p'$, $\rho = \rho_0 + \rho'$, $\vec{u} = \vec{u}'$ where $\frac{p'}{p_0} \ll 1$, $\frac{\rho'}{\rho_0} \ll 1$, and $\frac{u'}{c_0} \ll 1$,

Eqs 1.3.1 and 1.3.2 reduce to the following equations when second-order terms are neglected,

$$\frac{\partial \rho'}{\partial t} + \vec{\nabla} \cdot \rho_0 \vec{u}' = 0, \tag{1.5.1}$$

$$\frac{\partial \vec{u}'}{\partial t} + \left(\vec{u}' \cdot \vec{\nabla}\right) \vec{u}' + \frac{1}{\rho_0} \vec{\nabla} p' = 0. \tag{1.5.2}$$

For low frequency perturbations, the convection term is small compared to the non-steady term, and Eq 1.5.2 becomes

$$\frac{\partial \vec{u}'}{\partial t} + \frac{1}{\rho_0} \vec{\nabla} p' = 0. \tag{1.5.3}$$

Since $c_0^2 = \frac{p'}{\rho'}$, Eq 1.5.1 can be written as

$$\frac{\partial p'}{\partial t} + \rho_0 c_0^2 \vec{\nabla} \cdot \vec{u}' = 0, \tag{1.5.4}$$

and taking the divergence of Eq 1.5.3 and combining with Eq 1.5.4, we get

$$\frac{\partial^2 p'}{\partial t^2} - c_0^2 \nabla^2 p' = 0, \tag{1.5.5}$$

which is the linear wave equation in the variable p'. Defining a velocity potential ϕ such that $\vec{\nabla} \phi = \vec{u}'$, we may write Eq 1.5.3 as

$$\vec{\nabla} \frac{\partial \phi}{\partial t} + \vec{\nabla} \frac{p'}{\rho_0} = 0.$$

Hence

$$p' = -\rho_0 \frac{\partial \phi}{\partial t}. \tag{1.5.6}$$

Differentiating the above equation with respect to time,

$$\frac{\partial p'}{\partial t} = -\rho_0 \frac{\partial^2 \phi}{\partial t^2},$$

and replacing $\frac{\partial p'}{\partial t}$ in Eq 1.5.4 by the above equation yields

$$\phi_{tt} - c_0^2 \nabla^2 \phi = 0, \tag{1.5.7}$$

which is the linear wave equation in the perturbation velocity potential, ϕ. For one-dimensional planar, cylindrical, and spherical waves, Eq 1.5.7 can be written as

$$\phi_{tt} - \frac{c_0^2}{r^j} \frac{\partial}{\partial r} \left(r^j \phi_r\right) = 0, \tag{1.5.8}$$

where $j = 0, 1, 2$ for the planar, cylindrical, and spherical geometries, respectively.

For the planar case $j = 0$, the solution of Eq 1.5.8 is given by

$$\phi = f_1(r - c_0 t) + f_2(r + c_0 t), \tag{1.5.9}$$

corresponding to two sets of waves ($r \mp c_0 t = $ constant). The solution at any field point (r, t) is due to the superposition of perturbations carried there by the waves $r \mp c_0 t =$ constant. For spherical wave ($j = 2$), the solution of Eq 1.5.8 is given by

$$\phi(r,t) = \frac{1}{r} f_1(r - c_0 t) + \frac{1}{r} f_2(r + c_0 t). \tag{1.5.10}$$

In contrast to planar waves which propagate with constant amplitude and wave form, the amplitude of spherical waves varies with $\frac{1}{r}$ but the profile remains invariant.

The solution for cylindrical waves is more complicated. However since Eq 1.5.8 is linear, we can construct the solution by superposition of the solutions of spherical waves from elementary point source theory.

1.6 Acoustic Radiation from a Spherical Expanding Piston

As an example, consider a spherical piston expanding from the origin $r = 0$ at $t = 0$ with a constant velocity $u_p = \alpha c_0$, where $\alpha \ll 1$. The piston path is given by $r_p = \alpha c_0 t$. Since only outgoing diverging waves are involved, the solution is given by

$$\phi(r,t) = \frac{1}{r} f(r - c_0 t),$$

and the particle velocity and pressure can be expressed in terms of ϕ as

$$u' = \phi_r = \frac{1}{r} f' - \frac{1}{r^2} f,$$

$$p' = -\rho_0 \phi_t = \frac{\rho_0 c_0}{r} f'.$$

At the piston surface, $r = \alpha c_0 t$, $u' = u_p = \alpha c_0$,

$$u_p = \alpha c_0 = \frac{1}{r_p} f'\left(r_p - c_0 t\right) - \frac{1}{r_p^2} f\left(r_p - c_0 t\right).$$

Defining $\xi = r_p - c_0 t = c_0 t \left(\frac{r_p}{c_0 t} - 1\right) = c_0 t(\alpha - 1)$, we write

$$\alpha c_0 = \frac{1}{\alpha c_0 t} + f'(\xi) - \left(\frac{\alpha - 1}{\alpha \xi}\right)^2 f(\xi),$$

and solving for $f(\xi)$, we get

$$f(\xi) = \frac{\alpha^3 c_0}{\alpha^2 - 1} \xi^2 + C \xi^{\frac{\alpha - 1}{\alpha}}.$$

To evaluate the constant of integration, we apply the initial condition $t = 0$, $r_p = 0$, $\xi = 0$, $f(\xi) = 0$. Thus $C = 0$ and hence

$$f(\xi) = \frac{\alpha^3 c_0}{\alpha^2 - 1} \xi^2.$$

Since $\xi = r - c_0 t$, the above equation can be written as

$$f(r - c_0 t) = \frac{\alpha^3 c_0}{\alpha^2 - 1}(r - c_0 t)^2. \tag{1.6.1}$$

The particle velocity and pressure can be obtained as

$$u' = \frac{\alpha^3 c_0}{\alpha^2 - 1}\left(1 - \left(\frac{c_0 t}{r}\right)^2\right), \tag{1.6.2}$$

$$p' = -\rho_0 \phi_t = \frac{2\rho_0 c_0^2 \alpha^3}{\alpha^2 - 1}\left(1 - \frac{c_0 t}{r}\right). \tag{1.6.3}$$

At the piston surface $r = r_p = \alpha c_0 t$, Eq 1.6.2 gives

$$u' = \frac{\alpha^3 c_0}{\alpha^2 - 1}\left(1 - \left(\frac{c_0 t}{\alpha c_0 t}\right)^2\right) = \alpha c_0 = u_p.$$

The pressure at the piston face is given by Eq 1.6.3 as

$$p'_p = \frac{2\rho_0 c_0^2 \alpha^2}{\alpha + 1}. \tag{1.6.4}$$

The above results were obtained by Taylor (1946).

Acoustic radiation from a uniformly expanding piston starting initially at the center of symmetry has a self-similar solution also.

The acoustic equation is given by Eq 1.5.7 as

$$\phi_{tt} - c_0^2 \phi rr - \frac{j c_0^2 \phi_r}{r} = 0, \tag{1.6.5}$$

where the subscripts r, t refer to differentiation with respect to r and t, c_0 is the sound speed in the undisturbed medium, and $j = 1, 2$ for cylindrical and spherical geometries, respectively. We shall seek a similarity solution to Eq 1.6.5 of the form

$$\phi(r, t) = c_0^2 t F(\xi), \tag{1.6.6}$$

where $\xi = \frac{r}{c_0 t}$ is the similarity variable. (Here ϕ denotes the velocity potential.) Using Eq 1.6.6, the acoustic wave equation becomes

$$(\xi^2 - 1)F''(\xi) - \frac{jF'(\xi)}{\xi} = 0, \tag{1.6.7}$$

where prime denotes differentiation with respect to ξ. Defining $G(\xi) = F'(\xi)$, Eq 1.6.7 becomes

$$(\xi^2 - 1)G'(\xi) - \frac{jG(\xi)}{\xi} = 0, \tag{1.6.8}$$

which integrates to yield

$$G(\xi) = C\left(\frac{1 - \xi^2}{\xi^2}\right)^{\frac{j}{2}}, \tag{1.6.9}$$

where C is the constant of integration and can be evaluated using the boundary conditions at the piston, that is, $\xi = \xi_p = \frac{r_p}{c_0 t} = \alpha$, $u = \dot{r}_p = \alpha c_0$. Since $u = \phi_r = c_0 F'(\xi) = c_0 G(\xi)$ at the $\xi = \xi_p$, $u_p = c_0 G(\xi_p) = \alpha c_0$, $G(\xi_p) = \alpha$, the constant of integration can be obtained as

$$C = \alpha \left(\frac{\alpha^2}{1-\alpha^2} \right)^{\frac{j}{2}}.$$

The solution for $G(\xi)$ then becomes

$$G(\xi) = \alpha \left(\frac{\alpha^2}{1-\alpha^2} \right)^{\frac{j}{2}} \left(\frac{1-\xi^2}{\xi^2} \right)^{\frac{j}{2}}. \tag{1.6.10}$$

Since $u' = \phi_r = c_0 G(\xi)$, the solution for the spherical case where $j = 2$ gives

$$\frac{u'}{c_0} = \frac{\alpha^3}{1-\alpha^2} \left(\frac{1}{\xi^2} - 1 \right) = \frac{\alpha^3}{1-\alpha^2} \left(\left(\frac{c_0 t}{r} \right)^2 - 1 \right). \tag{1.6.11}$$

To obtain the pressure, we integrate $G(\xi)$ to obtain $F(\xi)$, that is,

$$F(\xi) = \int C \left(\frac{1-\xi^2}{\xi^2} \right)^{\frac{j}{2}} \mathrm{d}\xi,$$

and for $j = 2$, the above equation integrates to yield

$$F(\xi) = -C \left(\frac{1+\xi^2}{\xi} \right) + K,$$

where K is a constant of integration. From the boundary condition $\xi = 1$, $F(\xi) = 0$, we obtain $K = 2C$; thus

$$F(\xi) = \frac{-C(\xi-1)^2}{\xi} = \frac{\alpha^3}{\alpha^2 - 1} \frac{(\xi-1)^2}{\xi}. \tag{1.6.12}$$

The pressure perturbation is given by $p' = -\rho_0 \phi_t = -\rho_0 c_0^2 [F(\xi) - \xi F'(\xi)]$ and substituting the expressions for $F(\xi)$ and $F'(\xi) = G(\xi)$ into the above equation gives

$$\Delta p' = \frac{\rho_0 c_0^2 \alpha^3}{\alpha^2 - 1} \left(1 - \frac{1}{\xi} \right) = \frac{\rho_0 c_0^2 \alpha^3}{\alpha^2 - 1} \left(1 - \frac{c_0 t}{r} \right), \tag{1.6.13}$$

which is identical to the result obtained by Taylor (1946).

For the cylindrical case $j = 1$, Eq 1.6.10 gives

$$G(\xi) = F'(\xi) = \alpha \left(\frac{\alpha^2}{1-\alpha^2} \right)^{\frac{1}{2}} \left(\frac{1-\xi^2}{\xi^2} \right)^{\frac{1}{2}}, \tag{1.6.14}$$

and since $u' = \phi_r = c_0 F'(\xi)$, the velocity field is obtained as

$$u' = \alpha c_0 \left(\frac{\alpha^2}{1-\alpha^2} \right)^{\frac{1}{2}} \left(\frac{1 - \left(\frac{r}{c_0 t} \right)^2}{\left(\frac{r}{c_0 t} \right)^2} \right)^{\frac{1}{2}},$$

or

$$u' = \alpha c_0 \left(\frac{\alpha^2}{1-\alpha^2}\right)^{\frac{1}{2}} \left(\left(\frac{c_0 t}{r}\right)^2 - 1\right)^{\frac{1}{2}}. \tag{1.6.15}$$

To obtain $F(\xi)$, we integrate Eq 1.6.14 once more and obtain

$$F(\xi) = \alpha \left(\frac{\alpha^2}{1-\alpha^2}\right)^{\frac{1}{2}} \left[(1-\xi^2)^{\frac{1}{2}} - \ln\left(\frac{1+(1-\xi^2)^{\frac{1}{2}}}{\xi}\right)\right] + K.$$

To evaluate the constant of integration K, we use the boundary condition $\xi = 1$, $F(\xi) = 0$. Doing so, K is found to be 0 and the solution is thus

$$F(\xi) = \alpha \left(\frac{\alpha^2}{1-\alpha^2}\right)^{\frac{1}{2}} \left[(1-\xi^2)^{\frac{1}{2}} - \ln\left(\frac{1+(1-\xi^2)^{\frac{1}{2}}}{\xi}\right)\right]. \tag{1.6.16}$$

The pressure field can then be obtained since $p' = -\rho_0 \phi_t = \rho_0 c_0^2[\xi G(\xi) - F(\xi)]$. Substituting the solutions for $G(\xi)$ and $F(\xi)$ into the above equation gives

$$p' = \rho_0 c_0^2 \alpha \left(\frac{\alpha^2}{1-\alpha^2}\right)^{\frac{1}{2}} \ln\left[\frac{1+(1-\xi^2)^{\frac{1}{2}}}{\xi}\right]. \tag{1.6.17}$$

For the planar case ($j = 0$), the solution simply gives a constant velocity distribution from the front to the piston face.

1.7 Waves of Finite Amplitude

For waves of finite amplitude, the conservation equations (Euler's equations) need to be solved. For planar waves ($j = 0$), Eqs 1.3.12 and 1.3.13 can be written as

$$\frac{\partial c}{\partial t} + u\frac{\partial c}{\partial r} + \left(\frac{\gamma-1}{2}\right)c\frac{\partial u}{\partial r} = 0, \tag{1.7.1}$$

$$\frac{\partial u}{\partial t} + u\frac{\partial u}{\partial r} + \left(\frac{2c}{\gamma-1}\right)\frac{\partial c}{\partial r} = 0. \tag{1.7.2}$$

Riemann obtained a solution of the above equation for $c(u)$ for simple waves. Rearranging Eqs 1.7.1 and 1.7.2, we write

$$\frac{\left(\frac{\partial c}{\partial t}\right)_r}{\left(\frac{\partial c}{\partial r}\right)_t} = -\left(u + \frac{\gamma-1}{2}c\frac{\left(\frac{\partial u}{\partial r}\right)_t}{\left(\frac{\partial c}{\partial r}\right)_t}\right), \tag{1.7.3}$$

$$\frac{\left(\frac{\partial u}{\partial t}\right)_r}{\left(\frac{\partial u}{\partial r}\right)_t} = -\left(u + \frac{2}{\gamma-1}c\frac{\left(\frac{\partial c}{\partial r}\right)_t}{\left(\frac{\partial c}{\partial r}\right)_t}\right). \tag{1.7.4}$$

Using the cyclic rule, that is,

$$\left(\frac{\partial c}{\partial t}\right)_r \left(\frac{\partial t}{\partial r}\right)_c \left(\frac{\partial r}{\partial c}\right)_t = -1,$$

we obtain

$$\left(\frac{\partial c}{\partial t}\right)_r = \frac{-1}{\left(\frac{\partial t}{\partial r}\right)_c \left(\frac{\partial r}{\partial c}\right)_t} = -\frac{\left(\frac{\partial c}{\partial r}\right)_t}{\left(\frac{\partial t}{\partial r}\right)_c}.$$

Hence

$$\frac{\left(\frac{\partial c}{\partial t}\right)_r}{\left(\frac{\partial c}{\partial r}\right)_t} = -\left(\frac{\partial r}{\partial t}\right)_c. \tag{1.7.5}$$

Similarly, we write

$$\left(\frac{\partial u}{\partial t}\right)_r \left(\frac{\partial t}{\partial r}\right)_u \left(\frac{\partial r}{\partial u}\right)_t = -1,$$

and obtain

$$\frac{\left(\frac{\partial u}{\partial t}\right)_r}{\left(\frac{\partial u}{\partial r}\right)_t} = -\left(\frac{\partial r}{\partial t}\right)_u. \tag{1.7.6}$$

Since $u(r, t)$, we write

$$\begin{aligned}
\mathrm{d}u &= \left(\frac{\partial u}{\partial r}\right)_t \mathrm{d}r + \left(\frac{\partial u}{\partial t}\right)_r \mathrm{d}t, \\
&= \left(\frac{\partial u}{\partial r}\right)_t \left(\mathrm{d}r + \frac{\left(\frac{\partial u}{\partial t}\right)_r}{\left(\frac{\partial u}{\partial r}\right)_t}\mathrm{d}t\right), \\
&= \left(\frac{\partial u}{\partial r}\right)_t \left(\mathrm{d}r - \left(\frac{\partial r}{\partial t}\right)_u \mathrm{d}t\right).
\end{aligned}$$

Similarly, we obtain

$$\mathrm{d}c = \left(\frac{\partial c}{\partial r}\right)_t \left(\mathrm{d}r - \left(\frac{\partial r}{\partial t}\right)_c \mathrm{d}t\right).$$

If $c(u)$, $\left(\frac{\partial r}{\partial t}\right)_u = \left(\frac{\partial r}{\partial t}\right)_c$, we have

$$\frac{\mathrm{d}u}{\mathrm{d}c} = \frac{\left(\frac{\partial u}{\partial r}\right)_t}{\left(\frac{\partial c}{\partial r}\right)_t}, \tag{1.7.7}$$

and

$$\frac{\mathrm{d}c}{\mathrm{d}u} = \frac{\left(\frac{\partial c}{\partial r}\right)_t}{\left(\frac{\partial u}{\partial r}\right)_t}. \tag{1.7.8}$$

From Eqs 1.7.3 and 1.7.4, we obtain

$$u + \frac{\gamma - 1}{2} c \frac{\mathrm{d}u}{\mathrm{d}c} = u + \frac{2}{\gamma - 1} c \frac{\mathrm{d}c}{\mathrm{d}u}. \tag{1.7.9}$$

Solving for $\frac{\mathrm{d}u}{\mathrm{d}c}$, we get

$$\frac{\mathrm{d}u}{\mathrm{d}c} = \pm\frac{2}{\gamma - 1}, \tag{1.7.10}$$

which integrates to yield

$$u = \pm\left(\frac{2}{\gamma - 1}\right)c + \text{constant}.$$

If $u = 0$, $c = c_0$, the constant can be evaluated and the above equation can be written as

$$u = \pm\frac{2}{\gamma - 1}(c - c_0),$$

or

$$c = c_0 \pm \frac{\gamma - 1}{2}u. \tag{1.7.11}$$

Equations 1.7.9 and 1.7.10 give

$$u + \frac{\gamma - 1}{2}\frac{\mathrm{d}u}{\mathrm{d}c} = u \pm c,$$

and from Eqs 1.7.3 and 1.7.4, we obtain

$$\left(\frac{\partial r}{\partial t}\right)_{u,c} = u \pm c, \tag{1.7.12}$$

which integrates to yield

$$r = (u \pm c(u))t + f(u). \tag{1.7.13}$$

Using Eq 1.7.11, the above equation can also be written as

$$r = \left(\pm c_0 + \frac{\gamma + 1}{2}u\right)t + f(u), \tag{1.7.14}$$

where $f(u)$ is a constant of integration. The Riemann solution given above is the same as the solution using the method of characteristic given previously (Eq 1.4.17). Equation 1.7.12 gives the propagation of finite amplitude waves which propagate at the local sound speed $u \pm c$. For small amplitude waves where $u \simeq 0$, $c \simeq c_0$, Eq 1.7.12 reduces $\frac{\mathrm{d}r}{\mathrm{d}t} = \pm c_0$ corresponding to the acoustic solution.

1.8 The Piston Problem

To illustrate the solution for finite amplitude waves, consider the non-steady isentropic flow generated by a piston whose path $x_p = \pm\frac{at^2}{2}$. The $\pm$ sign corresponds to piston motion in the $\pm x$ direction, generating compression and expansion waves, respectively, as given in Fig. 1.2. The unsteady flow region is bounded by the wave front $x = c_0 t$ and the piston $x_p = \pm at^2/2$. Since only one set of waves is involved and they are propagating

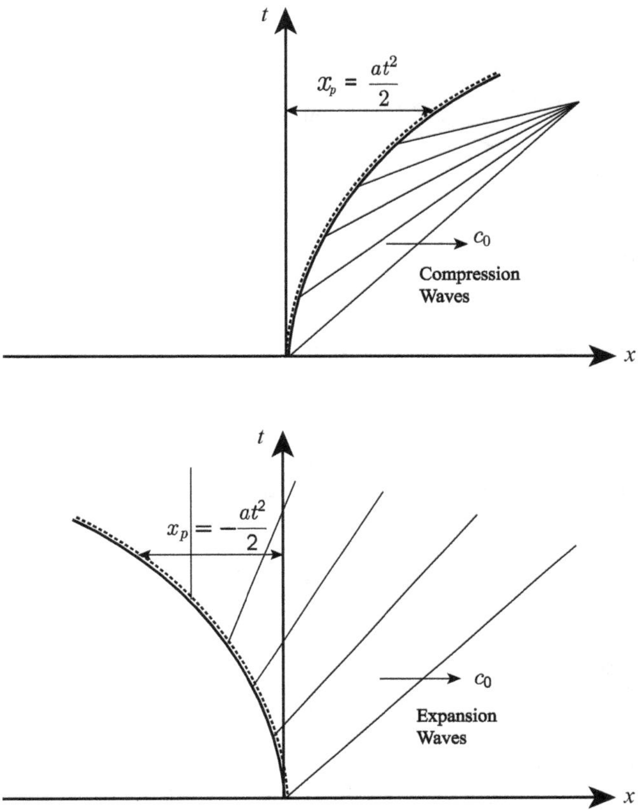

Figure 1.2 Compression Waves and Expansion Waves from Motion of a Piston

into a quiescent medium where $u = 0$, $c = c_0$, the flow is described by the simple wave solution (i.e., Eq 1.4.17).

$$x = \left(c_0 + \frac{\gamma + 1}{2}u\right)t + f(u),$$

or

$$f(u) = (x - c_0 t) - \left(\frac{\gamma + 1}{2}\right)ut. \tag{1.8.1}$$

To evaluate $f(u)$ we use the boundary condition at the piston surface $x_p = \pm at^2/2$, where $u = u_p = \pm at$. Thus we write Eq 1.8.1 as

$$\begin{aligned} f(u_p) &= \pm\frac{at^2}{2} - c_0 t - \frac{\gamma + 1}{2}at^2, \\ &= \mp\left(\frac{c_0 u_p}{a} + \frac{\gamma u_p^2}{2a}\right). \end{aligned} \tag{1.8.2}$$

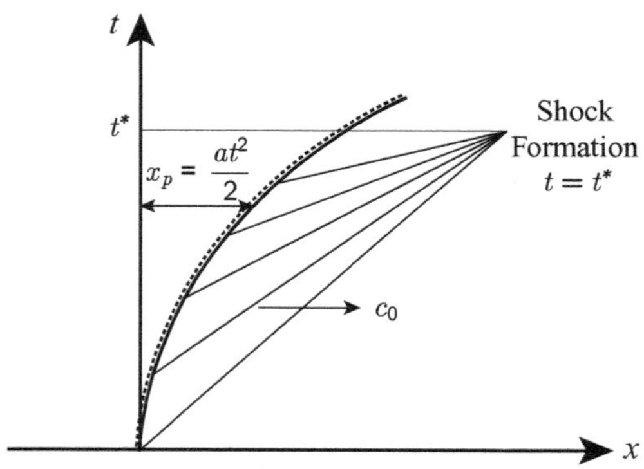

Figure 1.3 Formation of Shock from Converging Compression Waves

Substituting the above equation into Eq 1.8.1, we obtain the following quadratic equation for u, that is,

$$u^2 + \left(\frac{2c_0}{\gamma} + \frac{\gamma+1}{\gamma}(\mp at)\right)u \mp \frac{2a}{\gamma}(c_0 t - x) = 0.$$

Solving for u, we obtain

$$u = -\frac{1}{\gamma}\left(c_0 \mp \frac{\gamma+1}{2}at\right)\left[1 - \sqrt{1 \pm \frac{2\gamma a(c_0 t - x)}{\left(c_0 \mp \frac{\gamma+1}{2}at\right)^2}}\right]. \tag{1.8.3}$$

The top and bottom signs denote compression and expansion solutions, respectively.

For the compression solution corresponding to the piston path $x_p = +\frac{at^2}{2}$, the C^+ characteristics converge leading to shock formation when the characteristic coalesce at $t = t^*$. This is shown in Fig. 1.3. At the shock front, $\left(\frac{\partial u}{\partial x}\right)_{x=c_0 t^*} = \infty$, or $\left(\frac{\partial x}{\partial u}\right)_{x=c_0 t^*} = 0$, where t^* is when C^+ coalesces. Since $x = \left(c_0 + \frac{\gamma+1}{2}u\right)t + f(u)$, $\frac{\partial x}{\partial u} = f'(u) + \frac{\gamma+1}{2}t$ and equating $\frac{\partial x}{\partial u}$ to zero at $t = t^*$, we obtain

$$t^* = -\frac{2f'(u)\big|_{x=c_0 t}}{\gamma+1}. \tag{1.8.4}$$

Since $f(u) = -\left(\frac{c_0 u}{a} + \frac{\gamma u^2}{2a}\right)$ (i.e., Eq 1.8.2), we get $f'(u) = -\left(\frac{c_0}{a} + \frac{\gamma u}{a}\right)$ and evaluating at the front $x = c_0 t$, where $u = 0$, we obtain $f'(u)\big|_{x=c_0 t} = -\frac{c_0}{a}$. Thus Eq 1.8.4 gives

$$t^* = \frac{2c_0}{(\gamma+1)a}. \tag{1.8.5}$$

The distance for shock formation would be $x^* = c_0 t^* = \frac{2c_0^2}{(\gamma+1)a}$.

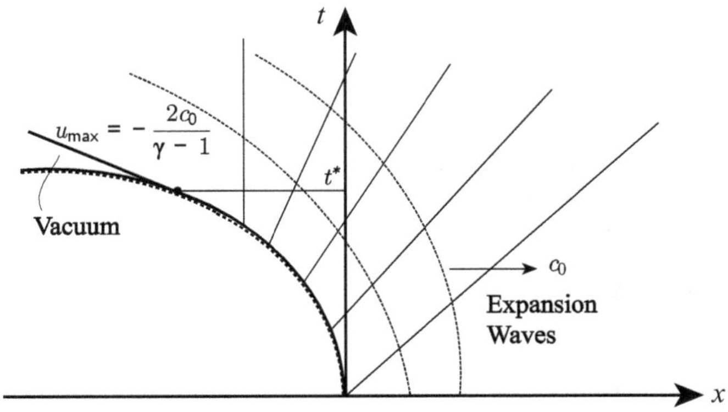

Figure 1.4 Region of Vacuum in the Expansion Solution

For the expansion solution corresponding to $x_p = -\frac{at^2}{2}$, there will be a maximum velocity u which occurs when the square root term in Eq 1.8.3 vanishes. Thus

$$u_{\max} = -\frac{1}{\gamma}\left(c_0 + \frac{\gamma + 1}{2}at\right). \tag{1.8.6}$$

Since the particle velocity at the piston surface equals the piston velocity, $u = u_p = -at$. The time t^* when $u = u_p = u_{\max}$ is thus $-at^* = -\frac{1}{\gamma}\left(c_0 + \frac{\gamma+1}{2}at^*\right)$ and solving for t^* gives

$$t^* = \frac{2c_0}{(\gamma - 1)a}. \tag{1.8.7}$$

Thus $u_{\max} = -at^* = -\frac{2c_0}{(\gamma-1)}$. For $t > t^*$, $u_p > -\frac{2c_0}{(\gamma-1)}$; however, the gas can no longer follow the piston and a vacuum region is obtained as shown in Fig. 1.4. $u_{\max}$ is referred to as the escape velocity. At the escape front, $p = 0$ and $c = 0$.

A particular case of interest is the centered rarefaction fan due to an impulsively accelerated piston at $t = 0$ and moving at a constant velocity for $t > 0$, that is, $x_p = -Ut$, $u_p = -U$. This is shown in Fig. 1.5. For the particular case where $f(u) = 0$, Eq 1.8.1 becomes

$$x = \left(c_0 + \frac{\gamma + 1}{2}u\right)t,$$

and rearranging, we write

$$\frac{u}{c_0} = \frac{2}{\gamma + 1}\left(\frac{x}{c_0 t} - 1\right). \tag{1.8.8}$$

At the head of the expansion fan, $\frac{x}{c_0 t} = 1$ and $\frac{u}{c_0} = 0$. At the tail of the expansion fan, $u = -U$ and

$$\frac{u}{c_0 t} = 1 - \left(\frac{\gamma + 1}{2}\right)\frac{U}{c_0}. \tag{1.8.9}$$

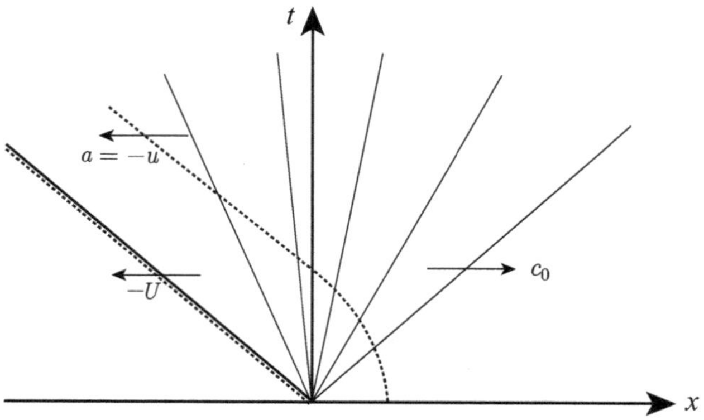

Figure 1.5 Centered Rarefaction Fan from an Impulsively Accelerated Piston

There will be a uniform region between the tail of the rarefaction fan and the piston. The sound speed can be obtained from Eq 1.7.11 as

$$\frac{c}{c_0} = 1 + \frac{\gamma - 1}{2}\frac{u}{c_0} = \frac{2}{\gamma + 1}\left(1 + \frac{\gamma - 1}{2}\frac{x}{c_0 t}\right). \tag{1.8.10}$$

From the sound speed, the pressure and density can be obtained from the isentropic relationships as

$$\frac{p}{p_0} = \left(\frac{c}{c_0}\right)^{\frac{2\gamma}{\gamma-1}} = \left[\frac{2}{\gamma + 1}\left(1 + \frac{\gamma - 1}{2}\frac{x}{c_0 t}\right)\right]^{\frac{2\gamma}{\gamma-1}}, \tag{1.8.11}$$

$$\frac{\rho}{\rho_0} = \left(\frac{c}{c_0}\right)^{\frac{2}{\gamma-1}} = \left[\frac{2}{\gamma + 1}\left(1 + \frac{\gamma - 1}{2}\frac{x}{c_0 t}\right)\right]^{\frac{2}{\gamma-1}}. \tag{1.8.12}$$

1.9 Shock Waves

A shock wave is a very thin compression wave that propagates at supersonic speed with respect to the gas ahead of it. Outside the extremely thin transition zone, transport effects can be ignored. If the changes in shock velocity are small over distances of the order of the shock thickness, the shock propagation can be considered as quasi-steady. Consider a shock wave propagating at velocity D into quiescent gas, shown in Fig. 1.6. The conservation equations across the shock are given by

$$\rho_0 D = \rho_1 (D - u_1), \tag{1.9.1}$$

$$p_0 + \rho_0 D^2 = p_1 + \rho_1 (D - u_1)^2, \tag{1.9.2}$$

$$h_0 + \frac{D^2}{2} = h_1 + \frac{(D - u_1)^2}{2}. \tag{1.9.3}$$

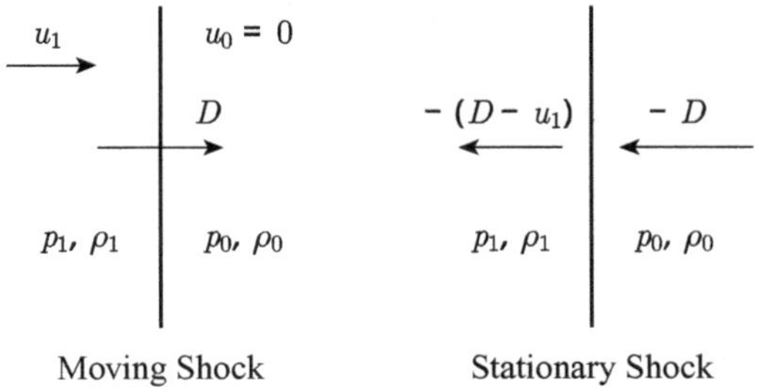

Figure 1.6 Shock Wave Propagating at Velocity D

From Eqs 1.9.1 and 1.9.2, we get

$$\frac{u_1}{D} = 1 - \frac{\rho_0}{\rho_1} = \frac{v_0 - v_1}{v_0}, \tag{1.9.4}$$

$$p_1 - p_0 = \rho_0 D u_1 = \frac{D^2}{v_0^2}(v_0 - v_1), \tag{1.9.5}$$

where $v = \frac{1}{\rho}$ is the specific volume. Equation 1.9.5 gives the shock velocity as

$$D = v_0 \sqrt{\frac{p_1 - p_0}{v_0 - v_1}}. \tag{1.9.6}$$

From Eq 1.8.1 we obtain

$$(D - u_1) = v_1 \sqrt{\frac{p_1 - p_0}{v_0 - v_1}}. \tag{1.9.7}$$

From Eqs 1.9.4 and 1.9.7, the particle velocity behind the shock can be obtained as

$$u_1 = \sqrt{(p_1 - p_0)(v_0 - v_1)}. \tag{1.9.8}$$

For D and u_1 to be real, we note that $p_1 > p_0$ and $v_0 > v_1$ (or $\rho_0 < \rho_1$) correspond to a compression shock. If $p_1 < p_0$, then $v_0 < v_1$ (or $\rho_0 > \rho_1$) for an expansion wave. Considering the compression shock solution, Eq 1.9.5 can also be written as

$$\frac{p_1}{p_0} = \left(1 + \frac{D^2}{p_0 v_0}\right) - \left(\frac{D^2}{p_0 v_0}\right)\frac{v_1}{v_0}, \tag{1.9.9}$$

corresponding to a straight line in the (pv) plane linking the initial $(p_0 v_0)$ to the shocked state $(p_1 v_1)$. Equation 1.9.9 is referred to as the Rayleigh line whose slope is proportional to the square of the shock velocity.

Eliminating the velocities in the energy equation (Eq 1.9.3) gives

$$h_1 - h_0 = \frac{1}{2}(p_1 - p_0)(v_0 + v_1). \tag{1.9.10}$$

In terms of the internal energy, the above equation can be written as

$$e_1 - e_0 = \frac{1}{2}(p_1 + p_0)(v_0 - v_1). \tag{1.9.11}$$

Equation 1.9.10 (and 1.9.11) is referred to as the shock Hugoniot curve giving the locus of shocked states for different shock velocities and a given initial state.

For a weak shock transition, we may expand $h(p, s)$ about the initial state (p_0, s_0), that is,

$$h(p,s) = h_0 + \left(\frac{\partial h}{\partial p}\right)(p_1 - p_0) + \frac{1}{2}\left(\frac{\partial^2 h}{\partial p^2}\right)(p_1 - p_0)^2 + \frac{1}{6}\left(\frac{\partial^3 h}{\partial p^3}\right)(p_1 - p_0)^3 + \cdots + \left(\frac{\partial h}{\partial s}\right)(s_1 - s_0) + \cdots \tag{1.9.12}$$

We need not consider terms higher than $(s_1 - s_0)$ since the entropy change is third order in the pressure change $(p_1 - p_0)$. From the thermodynamic relationship

$$dh = T ds + v dp = \left(\frac{\partial h}{\partial s}\right)_p ds + \left(\frac{\partial h}{\partial p}\right)_s dp,$$

we note

$$\left(\frac{\partial h}{\partial s}\right)_p = T,$$

$$\left(\frac{\partial h}{\partial p}\right)_s = v.$$

Thus we write Eq 1.9.12 as

$$h_1 - h_0 = T_0(s_1 - s_0) + v_0(p_1 - p_0) + \frac{1}{2}\left(\frac{\partial v}{\partial p}\right)_s (p_1 - p_0)^2 + \frac{1}{6}\left(\frac{\partial^2 v}{\partial p^2}\right)_s (p_1 - p_0)^3 + \cdots \tag{1.9.13}$$

Writing the Hugoniot equation (Eq 1.9.10) as

$$h_1 - h_0 = \frac{v_0}{2}(p_1 - p_0) + \frac{v_1}{2}(p_1 - p_0),$$

and expanding $v_i(p, s)$ about the initial state $v_0(p_0, s_0)$, we get

$$v_1 = v_0 + \left(\frac{\partial v}{\partial s}\right)_p (s_1 - s_0) + \cdots + \left(\frac{\partial v}{\partial p}\right)_s (p_1 - p_0) + \frac{1}{2}\left(\frac{\partial^2 v}{\partial p^2}\right)_s (p_1 - p_0)^2 + \frac{1}{6}\left(\frac{\partial^3 v}{\partial p^3}\right)_s (p_1 - p_0)^3 + \cdots$$

Substituting the above equation into the Hugoniot equation, we obtain

$$h_1 - h_0 = \frac{v_0}{2}(p_1 - p_0) + \frac{1}{2}\left[v_0(p_1 - p_0) + \left(\frac{\partial v}{\partial p}\right)_s (p_1 - p_0)^2 + \frac{1}{2}\left(\frac{\partial^2 v}{\partial p^2}\right)_s (p_1 - p_0)^3 + \cdots\right]. \tag{1.9.14}$$

We do not consider the term $\left(\frac{\partial v}{\partial s}\right)_p (s_1 - s_0)$ since it is already third order in $(p_1 - p_0)$. Equating Eqs 1.9.13 and 1.9.14, we obtain

$$T_0(s_1 - s_0) = +\frac{1}{12}\left(\frac{\partial^2 v}{\partial p^2}\right)_s (p_1 - p_0)^3 + \cdots \tag{1.9.15}$$

which shows that $(s_1 - s_0)$ is third order in $(p_1 - p_0)$ as mentioned earlier. From Eq 1.9.15, we note that $(s_1 - s_0) > 0$ if the right hand side is positive. Since $(p_1 - p_0) > 0$ for a compression shock, $(s_1 - s_0) > 0$ if $\left(\frac{\partial^2 v}{\partial p^2}\right)_s > 0$, since $c^2 = \left(\frac{\partial p}{\partial \rho}\right)_s$, we note $\left(\frac{\partial v}{\partial p}\right)_s = -\frac{v^2}{c^2}$. For a perfect gas with $\gamma =$ constant and $c^2 = \frac{\gamma p}{\rho} = \gamma p v$, $\left(\frac{\partial v}{\partial p}\right)_s = -\frac{v}{\gamma p}$ and $\left(\frac{\partial^2 v}{\partial p^2}\right)_s = \frac{\gamma+1}{\gamma}\frac{v}{p^2}$. Thus $\left(\frac{\partial^2 v}{\partial p^2}\right)_s > 0$ and $(s_1 - s_0) > 0$ for a compression shock in a perfect gas. Rarefaction shocks where $(p_1 - p_0) < 0$ would correspond to $(s_1 - s_0) < 0$ for a perfect gas. However, there are fluids where $\left(\frac{\partial^2 v}{\partial p^2}\right)_s < 0$ and rarefaction shocks can then be realized.

We can also demonstrate the behavior of the shock and particle velocity in the limit of a weak shock transition. Expanding $p_1(s, v)$ about the initial state $p_0(s_0, v_0)$, we write

$$p_1(s, v) = p_0 + \left(\frac{\partial p}{\partial v}\right)_s (v_1 - v_0) + \frac{1}{2}\left(\frac{\partial^2 p}{\partial v^2}\right)_s (v_1 - v_0)^2 + \frac{1}{6}\left(\frac{\partial^3 p}{\partial v^3}\right)_s (v_1 - v_0)^3 + \cdots + \left(\frac{\partial p}{\partial s}\right)_v (s_1 - s_0),$$

and substituting the above equation into Eq 1.9.6 we get

$$D^2 = -v_0^2\left[\left(\frac{\partial p}{\partial v}\right)_s + \frac{1}{2}\left(\frac{\partial^2 p}{\partial v^2}\right)_s (v_1 - v_0) + \cdots\right], \tag{1.9.16}$$

where the derivatives are evaluated at the initial state (p_0, v_0). For the lowest order in $(v_1 - v_0)$, the shock velocity is

$$D^2 = -v_0^2\left(\frac{\partial p}{\partial v}\right)_s = c_0^2, \tag{1.9.17}$$

indicating that a weak shock corresponds to an acoustic wave. We may write Eq 1.9.16 also as

$$D^2 = c_0^2 - \frac{1}{2}\left(\frac{\partial^2 p}{\partial v^2}\right)_s (v_1 - v_0) + \cdots \tag{1.9.18}$$

and since $v_1 - v_0$ and $\left(\frac{\partial^2 p}{\partial v^2}\right)_s > 0$ for normal fluids, we see that $D^2 > c_0^2$ or the shock propagates at supersonic speed relative to the gas ahead of it.

From Eq 1.9.8, we write

$$u_1^2 = (p_1 - p_0)(v_0 - v_1),$$

and substituting the expansion for $p_1(s, v)$ into the above equation, we obtain

$$u_1^2 = -\left[\left(\frac{\partial p}{\partial v}\right)_s (v_1 - v_0)^2 + \frac{1}{2}\left(\frac{\partial^2 p}{\partial v^2}\right)_s (v_1 - v_0)^3 + \cdots\right]. \tag{1.9.19}$$

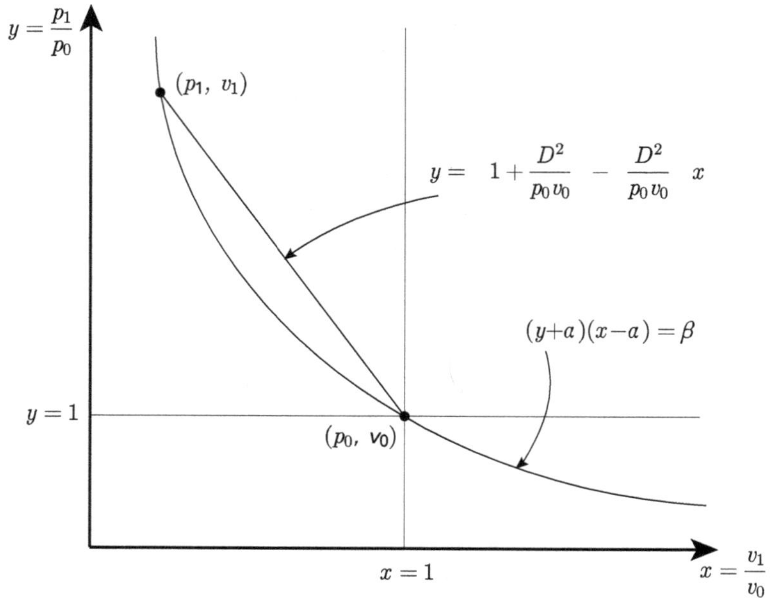

Figure 1.7 Shock Hugoniot and Rayleigh Line

Equation 1.9.18 shows D^2 is zeroth order in $(v_1 - v_0)$, whereas Eq 1.9.19 shows u_1^2 is second order in $(v_1 - v_0)$. Thus, the particle velocity change across a weak shock is very small.

For a perfect gas where the enthalpy is given by

$$h = \frac{\gamma p}{(\gamma - 1)\rho} = \frac{\gamma p v}{(\gamma - 1)}, \tag{1.9.20}$$

replacing the enthalpy in Eq 1.9.10 by the above equation, the Hugoniot equation can conveniently be expressed as a hyperbola of the form

$$(y + \alpha)(x - \alpha) = \beta, \tag{1.9.21}$$

where $y = \frac{p_1}{p_0}$, $x = \frac{v_1}{v_0}$, $\alpha = \frac{\gamma - 1}{\gamma + 1}$, and $\beta = 1 - \alpha^2 = \frac{4\gamma}{(\gamma+1)^2}$. Since the Hugoniot represents the locus of shock states and the Rayleigh line connects the upstream and downstream states across the shock, the transition across a shock corresponds to the intersection of the Rayleigh line with the shock Hugoniot as shown in Fig. 1.7. From Eq 1.9.1, we see that for strong shocks where $y \gg 1, x \to \alpha$ or

$$x = \frac{v_1}{v_0} = \frac{\rho_0}{\rho_1} \to \frac{\gamma - 1}{\gamma + 1},$$

giving the density ratio across a strong shock as

$$\frac{\rho_1}{\rho_0} = \frac{\gamma + 1}{\gamma - 1}. \tag{1.9.22}$$

It is interesting to note that the shocked density ratio approaches a finite limit for strong shocks. For isentropic compression, the density ratio approaches infinity when the pressure approaches infinity. The limiting density ratio can be understood if we consider the shock pressure can be achieved in two steps: isentropic compression to the shock density ratio followed by constant volume (density) heat addition to the shock pressure. Thus it is the dissipative heating in a shock transition that results in the density limit.

From Eq 1.9.4 the limiting ratio for the particle velocity can be obtained as

$$\frac{u_1}{D} = 1 - \frac{\rho_0}{\rho_1} = 1 - \frac{\gamma - 1}{\gamma + 1} = \frac{2}{\gamma + 1}. \tag{1.9.23}$$

Since γ is of the order of unity, we see that for strong shocks $u_1 \to D$. The limiting pressure ratio can be obtained from Eq 1.9.5. Neglecting p_0 as compared to p_1, we write

$$p_1 \cong \frac{D^2}{v_0^2}(v_0 - v_1) = \rho D^2 \left(1 - \frac{v_1}{v_0}\right) = \rho_0 D^2 \left(1 - \frac{\gamma - 1}{\gamma + 1}\right),$$

or

$$\frac{p_1}{\rho_0 D^2} = \frac{2}{\gamma + 1}. \tag{1.9.24}$$

Since $\gamma = \mathrm{O}[1]$, $D \cong u_1$ and the shocked pressure $p_1 \sim \rho_0 u_1^2$ corresponding to the dynamic pressure of the flow.

From the Hugoniot equation (Eq 1.9.11), the internal energy is given as

$$e_1 - e_0 = \frac{1}{2}(p_1 + p_0)(v_0 - v_1) \cong \frac{1}{2} p_1 v_0 \left(1 - \frac{v_1}{v_0}\right).$$

With $e_1 \gg e_0$, $p_1 \gg p_0$, the internal energy reduces to

$$e_1 \cong \frac{1}{2} p_1 v_0 \left(1 - \frac{v_1}{v_0}\right).$$

From Eq 1.9.8, we get

$$u_1^2 = (p_1 - p_0)(v_0 - v_1) \cong p_1 v_0 \left(1 - \frac{v_1}{v_0}\right),$$

where we have neglected p_0 as compared to p_1. Thus in the strong shock limit, the internal energy approaches the kinetic energy, $e_1 \cong \frac{u_1^2}{2}$.

The shocked state (i.e., ρ_1, p_1, u_1, etc.) can be expressed in terms of the initial state (ρ_0, p_0, etc.) and the shocked velocity (or Mach number, $M_s = \frac{D}{c_0}$). Using the Rayleigh line (Eq 1.9.9) to eliminate the pressure in the Hugoniot equation (Eq 1.9.21), we obtain a quadratic equation for the density ratio. Solving the quadratic equation and taking the appropriate root for a compression shock, we obtain

$$\frac{\rho_1}{\rho_0} = \frac{\gamma + 1}{(\gamma - 1) + 2\eta} = \frac{(\gamma + 1) M_s^2}{(\gamma - 1) M_s^2 + 2}, \tag{1.9.25}$$

where $\eta = \frac{1}{M_s^2}$. As $M_s \to \infty$, $\frac{\rho_1}{\rho_0} \to \frac{\gamma+1}{\gamma-1}$ in accordance with Eq 1.9.22. Using Eq 1.9.22, the particle velocity is obtained as

$$\frac{u_1}{D} = \frac{2(M_s^2 - 1)}{(\gamma + 1)M_s^2} = \frac{2}{\gamma + 1}(1 - \eta). \tag{1.9.26}$$

For strong shocks where $M_s \to \infty$, $\eta \to 0$, the velocity ratio becomes

$$\frac{u_1}{D} \to \frac{2}{\gamma + 1},$$

as obtained previously (i.e., Eq 1.9.23). Similarly, the pressure can be obtained as

$$\frac{p_1}{\rho_0 D^2} = \frac{2\gamma M_s^2 - (\gamma - 1)}{\gamma(\gamma + 1)M_s^2} = \frac{2}{\gamma + 1}\left(1 - \frac{\gamma - 1}{2\gamma}\eta\right), \tag{1.9.27}$$

which reduces to Eq 1.9.24 in the strong shock limit.

Equations 1.9.25–1.9.27 are referred to as the Rankine–Hugoniot relationships and are valid for shock waves in a perfect gas. However, the basic conservation laws and all the subsequent discussions prior to the use of the equation of state for a perfect gas are valid for an arbitrary medium.

1.10 Detonation and Deflagration Waves

Detonation and deflagration waves are combustion waves across which energy is released from chemical reactions. Although these combustion waves are orders of magnitudes thicker than a shock wave, they are nevertheless small compared to macroscopic length scales of the problem in general. Thus, they can be approximated as thin fronts across which the quasi-steady one-dimensional conservation equations apply. Although there is a change in composition and fluid properties across the combustion front, we shall assume a perfect gas with constant γ upstream and downstream of the wave. Chemical reactions are treated as external energy addition to the gas across the front.

The conservation equations of mass and momentum are the same as those for a normal shock given previously (i.e., Eqs 1.9.1 and 1.9.2). However, the energy equation has to be modified to include the chemical energy release, that is,

$$h_0 + Q + \frac{D^2}{2} = h_1 + \frac{(D - u_1)^2}{2}, \tag{1.10.1}$$

where Q is the chemical energy released per unit mass across the combustion wave. The equations derived from the conservation of mass and momentum (i.e., Eqs 1.9.4–1.9.9) across a shock wave apply to a combustion wave. However, the Hugoniot equation is now given by

$$h_1 - (h_0 + Q) = \frac{1}{2}(p_1 - p_0)(v_0 + v_1), \tag{1.10.2}$$

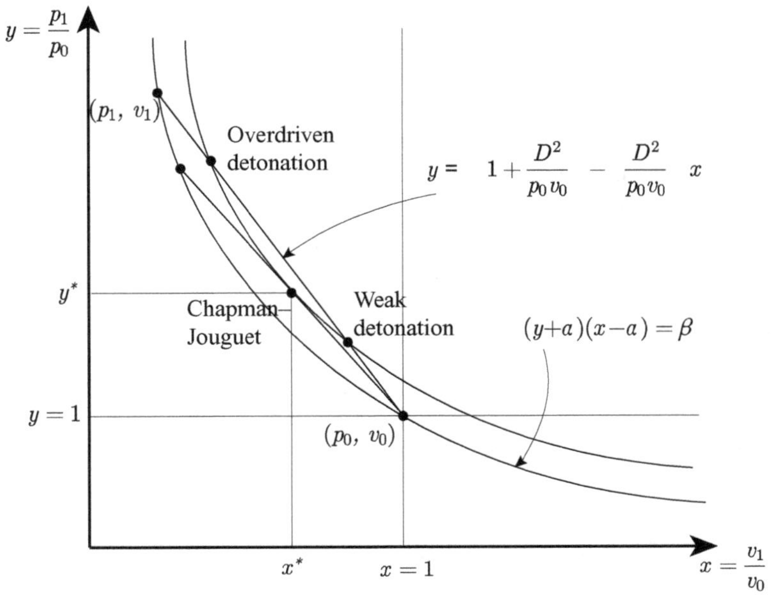

Figure 1.8 Overdriven, Chapman–Jouguet, and Weak Detonations

or in terms of the internal energy, we write

$$e_1 - (e_0 + Q) = \frac{1}{2}(p_1 + p_0)(v_0 - v_1). \tag{1.10.3}$$

For a perfect gas with constant γ, where the enthalpy and internal energy are given by

$$h = \frac{\gamma p}{(\gamma - 1)\rho},$$

$$e = \frac{p}{(\gamma - 1)\rho},$$

Eq 1.9.2 (or 1.10.2) can be written as

$$(y + \alpha)(x - \alpha) = \beta, \tag{1.10.4}$$

where $\alpha = \frac{\gamma-1}{\gamma+1}$, $\beta = 1 - \alpha^2 + \frac{2Q}{c_0^2}$, and $c_0^2 = \gamma p_0 v_0$ is the sound speed in the gas upstream of the front. Unlike a shock wave, the initial state (p_0, v_0) or $y = 1, x = 1$ does not lie on the Hugoniot curve given by Eq 1.10.4. Thus the Rayleigh line (Eq 1.9.9)

$$y = \left(1 + \frac{D^2}{p_0 v_0}\right) - (D^2 p_0 v_0)\, x,$$

will intersect the Hugoniot curve at two points, that is, for a given front velocity, there corresponds two solutions for the downstream state: the upper overdriven detonation solution and the lower weak detonation solution (Fig. 1.8).

The initial state lies on the shock Hugoniot curve where $Q = 0$. The two solutions coalesce when the Rayleigh line is tangent to the Hugoniot curve. The tangency solution is the minimum velocity solution referred to as the Chapman–Jouguet detonation solution. There is no solution below the Chapman–Jouguet velocity as the Rayleign line no longer intersects the Hugoniot curve.

For the Chapman–Jouguet solution, the slopes of the Rayleign line and Hugoniot curve are the same, thus

$$\left(\frac{\mathrm{d}y}{\mathrm{d}x}\right)_R^* = \left(\frac{\mathrm{d}y}{\mathrm{d}x}\right)_H^*,$$

and from Eqs 1.9.9 and 1.10.4, we obtain

$$-\left(\frac{y^* - 1}{1 - x^*}\right) = -\left(\frac{y^* + \alpha}{x^* - \alpha}\right),$$

where we denote y^*, x^* as the Chapman–Jouguet state. Solving for y^* from the above equation gives

$$y^* = \frac{-x^*(1-\alpha)}{1+\alpha-2x^*} = \frac{-x^*}{\gamma - (\gamma+1)x^*}. \tag{1.10.5}$$

Using the above equation, we evaluate the slope of the Rayleigh line to be

$$\left(\frac{\mathrm{d}y}{\mathrm{d}x}\right)_R^* = -\left(\frac{y^* - 1}{1 - x^*}\right) = \frac{\gamma}{\gamma - (\gamma+1)x^*}. \tag{1.10.6}$$

Multiplying Eq 1.10.6 by x^* in the numerator and denominator and using Eq 1.9.5, we get

$$\left(\frac{\mathrm{d}y}{\mathrm{d}x}\right)_R^* = \frac{\gamma x^*}{x^*\left[\gamma - (\gamma+1)x^*\right]} = \frac{-\gamma y^*}{x^*} = \left(\frac{\mathrm{d}y}{\mathrm{d}x}\right)_H^*. \tag{1.10.7}$$

The equation for an isentropic flow is given by $pv^\gamma = \text{constant}$, thus the slope is

$$\left(\frac{\mathrm{d}y}{\mathrm{d}x}\right)_s = -\frac{\gamma y}{x}. \tag{1.10.8}$$

From Eq 1.10.7, we see that at the tangency point, the slopes of the Rayleigh line, Hugoniot, and the isentrope are all equal

$$\left(\frac{\mathrm{d}y}{\mathrm{d}x}\right)_R^* = \left(\frac{\mathrm{d}y}{\mathrm{d}x}\right)_H^* = \left(\frac{\mathrm{d}y}{\mathrm{d}x}\right)_s^*. \tag{1.10.9}$$

The sound speed is related to the slope of the isentrope by

$$c^2 = \left(\frac{\partial p}{\partial \rho}\right)_s = -v^2\left(\frac{\partial p}{\partial v}\right)_s = -p_0 v_0 x^2 \left(\frac{\mathrm{d}y}{\mathrm{d}x}\right)_s,$$

or

$$\left(\frac{\mathrm{d}y}{\mathrm{d}x}\right)_s = \frac{-c^2}{p_0 v_0 x^2}. \tag{1.10.10}$$

From Eq 1.9.7, we write

$$(D - u_1^*)^2 = v_1^{*2}\frac{p_1^* - p_0}{v_0 - v_1} = p_0 v_0 x^{*2}\frac{y^* - 1}{1 - x^*},$$

and

$$\left(\frac{D-u_1^*}{c^*}\right) = M_1^{*2} = 1. \qquad (1.10.11)$$

Since $(D - u_1)$ is the particle velocity of the downstream flow relative to the wave, Eq 1.10.11 indicates that for Chapman–Jouguet detonations, the flow is sonic relative to the wave. For the overdriven detonation (upper intersection point) and the weak detonation (lower intersection point) solutions, it can be shown that $M_1 < 1$ and $M_1 > 1$ for overdriven and weak detonations, respectively. The overdriven detonation solution requires the boundary condition in the products downstream of the wave to be specified since the flow is subsonic. The weak detonation solution requires some additional information on the propagation mechanism of the detonation to be given. For freely propagating unsupported detonations, only Chapman–Jouguet detonations can be realized. Consideration of the detailed structure of the detonation may lead to a weak detonation solution. But on the basis of just the conservation laws across the wave, it is not possible to determine the weak detonation solution.

Similar to the procedure used for non-reacting shock waves, the Rankine–Hugoniot relationships across a detonation wave can be obtained using the Rayleigh line and the Hugoniot curve. Solving the quadratic equation for the density ratio obtained by eliminating the pressure y in the Hugoniot equation (Eq 1.10.4) using the Rayleigh line (Eq 1.10.9), we obtain

$$x = \frac{\rho_0}{\rho_1} = \frac{\gamma + \eta \pm S}{\gamma + 1}, \qquad (1.10.12)$$

$$\frac{u_1}{D} = \frac{1 - \eta \mp S}{\gamma + 1}, \qquad (1.10.13)$$

$$\frac{p_1}{\rho_0 D^2} = \frac{\gamma + \eta \mp \gamma S}{\gamma(\gamma + 1)}, \qquad (1.10.14)$$

where

$$S = \sqrt{(1-\eta)^2 - K\eta},$$

$$K = 2(\gamma^2 - 1)\frac{Q}{c_0^2},$$

$$\eta = \frac{1}{M_s^2} = \frac{c_0^2}{D^2}.$$

The $\pm$ signs in Eqs 1.10.12–1.10.14 refer to the two solutions of overdriven and weak detonations. For the tangency Chapman–Jouguet solution, we have a unique solution and $S = 0$ giving the Chapman–Jouguet velocity as

$$\eta_{CJ} = \frac{1}{M_{CJ}^2} = \left(1 + \frac{K}{2}\right) - \sqrt{K\left(\gamma + \frac{K}{4}\right)}.$$

In general, the Chapman–Jouguet Mach number $M_{CJ} \gg 1$ and $\eta_{CJ} \ll 1$ and neglecting η as compared to unity (note that K is sufficiently large so that $K\eta$ is not small compared to unity), we write

$$S = \sqrt{1 - K\eta},$$

and for Chapman–Jouguet detonations, $S = 0$ and solving for M_{CJ} yields

$$M_{CJ} = \sqrt{2(\gamma^2 - 1)\frac{Q}{c_0^2}},$$

that is, the detonation Mach number (or velocity) is proportional to the square root of the chemical energy release.

The above discussions are concerned with the upper detonation branch of the Hugoniot curve. Deflagration waves correspond to solutions on the lower branch of the Hugoniot curve. Deflagrations are subsonic waves and the boundary conditions play an important role apart from the mechanisms that govern the propagation of the deflagration itself. The deflagration solution cannot be obtained based on the gas dynamic analysis alone as the gas ahead of the combustion front is influenced by the rear boundary conditions. In general, a precursor shock wave is formed ahead of the deflagration and the upstream state of the deflagration front corresponds to the shocked state of the precursor shock. Solutions for deflagration require the simultaneous solution for the precursor shock and the deflagration front satisfying the prescribed boundary condition of the problem.

2 Weak Shock Theory

2.1 Introduction

The theoretical description of non-steady shock propagation is difficult in general as it involves the integration of the non-linear conservation equations. However, if the shock wave is weak, the entropy change across the shock is small and the shock can be treated as an isentropic compression wave. For a weak shock propagating into a uniform or quiescent medium, the flow behind it is a simple wave flow which can be determined by initial or boundary conditions. The shock motion can then be obtained by fitting the shock to the known simple wave flow behind it. The weak shock theories described in this chapter are of this nature. In Chandrasekhar's (1943) method, the characteristic equation is first transformed into the kinematic wave equation which has a particular solution corresponding to a linear distribution of flow variables behind the shock. The decay of the shock is then obtained by fitting the shock to the solution of the flow behind it. In the method of Friedrichs (1948), a simple wave flow is assumed behind the shock. The shock motion is then determined by matching the shock strength to give the same state as that carried by the characteristic that intersects it. The weak shock theory of Whitham (1956) is also very similar in nature to that of Freidrichs' method, except Whitham used a modified acoustic solution for the flow behind the shock. By modified acoustic theory, the linear acoustic solution is "non-linearized" by replacing the acoustic trajectories by the characteristics. In so doing, the distortion of the wave front can now be realized, whereas linear acoustic theory would give a wave front of permanent form. Oswatitsch (1956) also gave a theory for weak shock propagation. He assumed, as in Chandrasekhar's solution, a linear profile behind the shock and obtained a perturbation solution of the flow equations in the neighborhood of the sonic point ($u = 0$, $c = c_0$) of the triangular shock profile. Since the profile is linear, the solution at the sonic point can be extended to the shock front. The shock strength is then matched to the solution behind it. The principal approximations used in all these weak shock theories are essentially similar. The difference lies in the description of the flow behind it and the method used to match the shock to the flow.

2.2 Properties of Weak Shocks

In terms of the shock overpressure $y = \frac{\Delta p_1}{p_0}$, the Rankine–Hugoniot equations across a normal shock are given as

$$\frac{D}{c_0} = M_s = \left(1 + \frac{\gamma+1}{2\gamma}y\right)^{\frac{1}{2}}, \tag{2.2.1}$$

$$\frac{u_1}{c_0} = \frac{y}{\gamma\left(1 + \frac{\gamma+1}{2\gamma}y\right)}, \tag{2.2.2}$$

$$\frac{\rho_1}{\rho_0} = \frac{2\gamma + (\gamma+1)y}{2\gamma + (\gamma-1)y}, \tag{2.2.3}$$

$$\frac{c_1}{c_0} = \left\{\frac{(y+1)\left(2\gamma + (\gamma-1)y\right)}{2\gamma + (\gamma+1)y}\right\}^{\frac{1}{2}}, \tag{2.2.4}$$

$$\frac{\Delta s}{C_v} = \ln \frac{(1+y)\left(1 + \frac{\gamma-1}{2\gamma}y\right)^{\gamma}}{\left(1 + \frac{\gamma+1}{2\gamma}y\right)^{\gamma}}. \tag{2.2.5}$$

For weak shocks where $y \ll 1$, the above equations can be expanded in powers of y and we obtain

$$\frac{D}{c_0} = M_s = 1 + \frac{\gamma+1}{4\gamma}y - \frac{(\gamma+1)^2}{32\gamma^2}y^2 + \cdots \tag{2.2.6}$$

$$\frac{u_1}{c_0} = \frac{y}{\gamma} - \frac{\gamma+1}{4\gamma^2}y^2 + \frac{3(\gamma+1)^2}{32\gamma^3}y^3 + \cdots \tag{2.2.7}$$

$$\frac{\rho_1}{\rho_0} = 1 + \frac{y}{\gamma} - \frac{\gamma-1}{2\gamma^2}y^2 + \cdots \tag{2.2.8}$$

$$\frac{c_1}{c_0} = 1 + \frac{\gamma-1}{2\gamma}y - \frac{\gamma^2-1}{8\gamma^2}y^2 + \frac{(\gamma-1)(\gamma+1)^2}{16\gamma^3}y^3 + \cdots \tag{2.2.9}$$

$$\frac{\Delta s}{C_v} = \frac{\gamma^2-1}{12\gamma^2}y^3 + \frac{\gamma^2-1}{8\gamma^2}y^4 + \cdots \tag{2.2.10}$$

Up to first order in y, Eq 2.2.8 can be written as

$$\frac{\rho_1}{\rho_0} - 1 = \frac{y}{\gamma} = \frac{1}{\gamma}\left(\frac{\Delta p_1}{p_0}\right) = \Delta\left(\frac{\rho_1}{\rho_0}\right).$$

Dropping the subscript, we write

$$\frac{\Delta p}{\Delta \rho} = \frac{\gamma p_0}{\rho_0} = c_0^2,$$

which is just the expression for an acoustic wave. If we consider up to first order in y in Eqs 2.2.6 and 2.2.7, we obtain

$$\frac{D}{c_0} = 1 + \frac{\gamma+1}{4\gamma} y = 1 + \frac{\gamma+1}{4\gamma}\frac{u_1}{c_0},$$

and Eq 2.2.9 gives

$$\frac{c_1}{c_0} = 1 + \frac{\gamma-1}{2\gamma} y = 1 + \frac{\gamma-1}{2\gamma}\frac{u_1}{c_0}.$$

Combining the above expressions give

$$D = \frac{u_1 + c_1 + c_0}{2}, \tag{2.2.11}$$

which indicates that the shock velocity is given by the average of the slopes of the characteristics on both sides of the shock front. Equation 2.2.11 is a useful approximation for weak shocks and is generally referred to as "Pfriem's formula."

Equation 2.2.10 shows that the entropy increase across the shock is third order in y and higher. Thus even for moderate strength shocks, the entropy change across the shock is small, and hence weak shocks can be approximated as isentropic compression waves. For an isentropic process,

$$\frac{c_1}{c_0} = \left(\frac{T_1}{T_0}\right)^{\frac{1}{2}} = \left(\frac{p_1}{p_0}\right)^{\frac{\gamma-1}{2\gamma}} = \left(1 + \frac{\Delta p_1}{p_0}\right)^{\frac{\gamma-1}{2\gamma}} = 1 + \frac{\gamma-1}{2\gamma} y + \cdots$$

which essentially corresponds to Eq 2.2.9 of the Rankine–Hugoniot relationships.

A weak shock propagating into quiescent gas is shown in Fig. 2.1.

We note that along a C^- characteristic, the Riemann invariant $J^- =$ constant. Thus

$$J^- = -\frac{2}{\gamma-1} c_0 = u_1 - \frac{2}{\gamma-1} c_1 = u - \frac{2}{\gamma-1} c,$$

and the change in the Riemann invariant across the shock is

$$\frac{\Delta J^-}{c_0} = \left(\frac{u_1}{c_0} - \frac{2}{\gamma-1}\frac{c_1}{c_0}\right) - \left(-\frac{2}{\gamma-1}\right) = \frac{u_1}{c_0} + \frac{2}{\gamma-1}\left(1 - \frac{c_1}{c_0}\right).$$

Using Eqs 2.2.7 and 2.2.9, we get

$$\frac{\Delta J^-}{c_0} = -\frac{(\gamma+1)^2}{32\gamma^3} y^3 + \cdots$$

Thus up to second order in y, J^- is constant across the shock. Writing

$$J^- = u - \frac{2}{\gamma-1} c = -\frac{2}{\gamma-1} c_0,$$

we obtain the relationship between c and u as

$$c = c_0 + \frac{\gamma-1}{2} u, \tag{2.2.12}$$

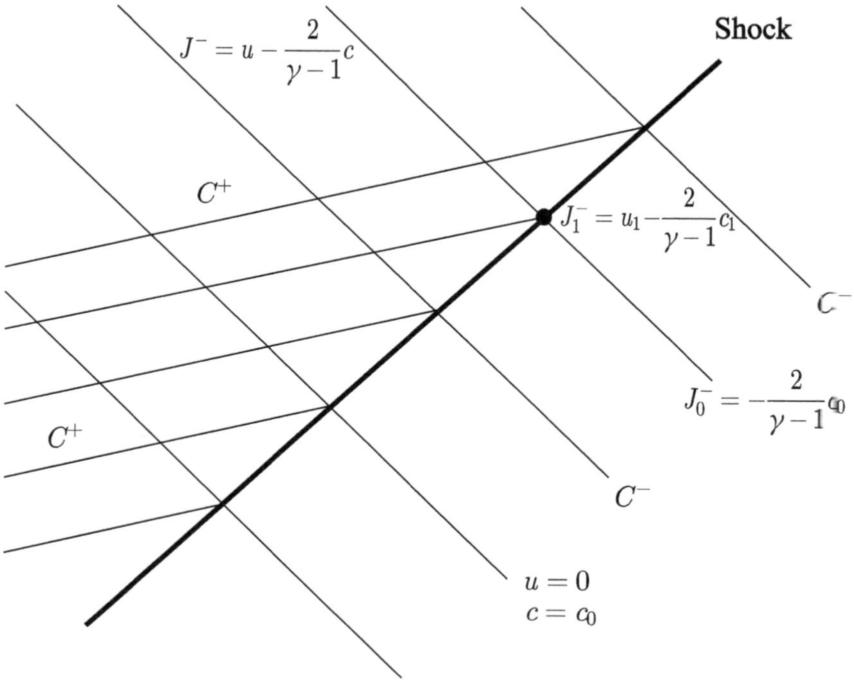

Figure 2.1 Weak Shock Propagating in a Quiescent Gas

for the flow behind the shock. Equation 2.2.12 for $c(u)$ defines a simple wave flow. For a C^+ in the simple wave flow, $J^+ = u + \frac{2}{\gamma-1}c =$ constant and using Eq 2.2.12, we obtain

$$J^+ = \frac{2}{\gamma-1}c_0\left(1 + (\gamma-1)\frac{u}{c_0}\right) = \frac{2c_0}{\gamma-1}\left(\frac{2c}{c_0} - 1\right). \tag{2.2.13}$$

Thus, each C^+ in the simple wave region behind the shock is defined by a value of u or equivalently a value of c. The slope of a C^+ is given by

$$\frac{dx}{dt} = u + c = c_0 + \frac{\gamma+1}{2}u = \frac{2}{\gamma-1}\left(\frac{\gamma+1}{2}c - c_0\right). \tag{2.2.14}$$

The slope of a C^+ is thus defined by the initial value of u (or c) in the simple wave region behind the shock. Integrating Eq 2.2.14, the trajectory of a C^+ is given by

$$x = \left(c_0 + \frac{\gamma+1}{2}u\right)t + f(u),$$

$$= \frac{2}{\gamma-1}\left(\frac{\gamma+1}{2}c - c_0\right)t + f(c),$$

where $f(u)$ or $f(c)$ are constants of integration to be determined from initial or boundary conditions.

The simple wave flow behind the shock can be determined independently of the shock motion. If the influence of the reflected characteristics from the shock can be neglected, then the shock motion can be determined by fitting the shock to a known flow field behind it. The decoupling of the shock motion from the flow behind it eliminates the need for the simultaneous solution of the flow and the shock motion. This greatly simplifies the solution of the shock propagation.

2.3 Chandrasekhar's Solution

Chandrasekhar (1943) obtained a solution for the decay of a planar weak shock pulse. Using the simple wave equations (Eqs 2.2.13 and 2.2.14), the characteristic equation in planar geometry, that is,

$$\left[\frac{\partial}{\partial t} + (u+c)\frac{\partial}{\partial x}\right]\left(u + \frac{2}{\gamma-1}c\right) = 0,$$

can be written as

$$\left[\frac{\partial}{\partial t} + \left(\frac{\gamma+1}{\gamma-1}c - \frac{2}{\gamma-1}c_0\right)\frac{\partial}{\partial x}\right]\left(\frac{2}{\gamma-1}(2c-c_0)\right) = 0.$$

Defining $\bar{c} = \frac{c}{c_0}$ and $\bar{t} = c_0 t$, the above equation becomes

$$\frac{\partial \bar{c}}{\partial \bar{t}} + \left(\frac{\gamma+1}{\gamma-1}\bar{c} - \frac{2}{\gamma-1}\right)\frac{\partial \bar{c}}{\partial x} = 0. \tag{2.3.1}$$

In terms of a new variable ϕ defined as

$$\phi = \frac{\gamma+1}{\gamma-1}\bar{c} - \frac{2}{\gamma-1}, \tag{2.3.2}$$

Eq 2.3.1 reduces to the form

$$\frac{\partial \phi}{\partial t} + \phi\frac{\partial \phi}{\partial x} = 0. \tag{2.3.3}$$

Equation 2.3.3 is generally referred to as the kinematic wave equation and has a particular solution given by

$$\phi = \frac{1+Ax}{B+A\bar{t}}. \tag{2.3.4}$$

Rescaling the time by defining

$$\tau = \bar{t} - \left(\frac{B-1}{A}\right), \tag{2.3.5}$$

the solution can be conveniently written as

$$\phi = \frac{1+Ax}{1+A\tau}. \tag{2.3.6}$$

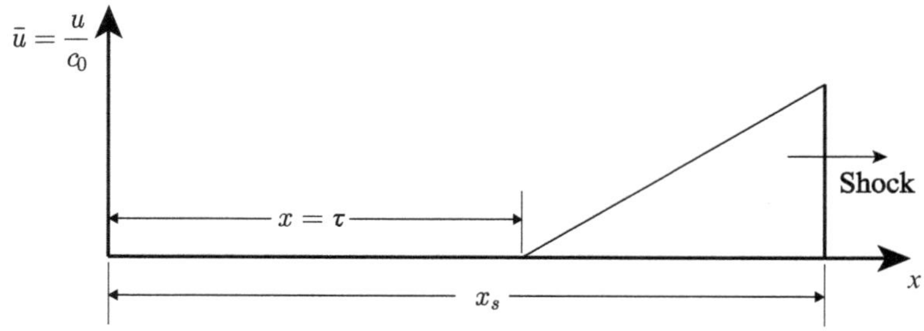

Figure 2.2 Particle Velocity Profile Behind a Weak Shock

From the definition of ϕ (Eq 2.3.2) and the simple wave relationship Eq 2.2.12, the solutions for $\bar{c}$ and $\bar{u}$ can be obtained as

$$\bar{c} = \frac{c}{c_0} = \left(\frac{\gamma - 1}{\gamma + 1}\right)\left(\frac{1 + Ax}{1 + A\tau} + \frac{2}{\gamma - 1}\right), \tag{2.3.7}$$

$$\bar{u} = \frac{u}{c_0} = \left(\frac{2}{\gamma + 1}\right)\left(\frac{1 + Ax}{1 + A\tau} - 1\right). \tag{2.3.8}$$

From Eqs 2.3.7 and 2.3.8, we note that the velocity and sound speed profiles behind the shock are linear for a given instant of time. The particle velocity profile is illustrated in Fig. 2.2.

Solving Eq 2.3.8 for $\bar{u} = \bar{u}_1$ at the shock front $x = x_s$ yields

$$x_s = \frac{1}{A}\left[(1 + A\tau)\left(1 + \frac{\gamma + 1}{2}\bar{u}_1\right) - 1\right],$$

and differentiating the above equation, we obtain

$$\begin{aligned} \dot{x_s} &= \frac{dx_s}{dt} = c_0\frac{dx_s}{d\tau} = c_0 M_s, \\ &= c_0\left[\left(1 + \frac{\gamma + 1}{2}\bar{u}_1\right) + \left(\frac{1 + A\tau}{A}\right)\frac{\gamma + 1}{2}\frac{d\bar{u}_1}{d\tau}\right]. \end{aligned} \tag{2.3.9}$$

The Rankine–Hugoniot relationships give

$$\bar{u}_1 = \frac{u_1}{c_0} = \frac{2}{\gamma + 1}\left(M_s - \frac{1}{M_s}\right),$$

and differentiating the above equation gives

$$\frac{d\bar{u}_1}{d\tau} = \left(\frac{2}{\gamma + 1}\right)\left(1 + \frac{1}{M_s^2}\right)\frac{dM_s}{d\tau}.$$

Substituting the above expression for $\frac{d\bar{u}_1}{d\tau}$ into Eq 2.3.9 and re-arranging yields

$$\frac{1 + M_s^2}{(1 - M_s)M_s}dM_s = \frac{Ad\tau}{1 + A\tau},$$

which can be written as

$$\frac{2}{1 - M_s}\mathrm{d}M_s + \frac{\mathrm{d}M_s}{M_s} - \mathrm{d}M_s = \mathrm{d}\ln(1 + A\tau).$$

The above equation can readily be integrated to give

$$\left(\frac{1 - M_s^0}{M_s^0}\right)^2 \frac{M_s}{(1 - M_s)^2} \exp(M_s^0 - M_s) = 1 + A\tau. \tag{2.3.10}$$

In the above solution, we have used the initial condition $M_s = M_s^0$ when $\tau = \tau^0$. For weak shocks where $M_s \simeq 1$ and $(M_s - 1) \ll 1$, Eq 2.3.10 reduces to

$$(M_s - 1) \simeq \frac{M_s^0 - 1}{(1 + A\tau)^{1/2}}. \tag{2.3.11}$$

The approximate form above is valid up to second order in $(M_s - 1)$. The Rankine–Hugoniot equations give the shock overpressure as

$$y = \Delta\left(\frac{p_1}{p_0}\right) = \frac{4\gamma}{\gamma + 1}(M_s - 1).$$

Using Eq 2.3.11, we write

$$\Delta\left(\frac{p_1}{p_0}\right) = \frac{4\gamma}{\gamma + 1}\frac{M_s^0 - 1}{(1 + A\tau)^{1/2}}.$$

For large times, we observe from the above equation that the weak shock pulse gives $\Delta(\frac{p_1}{p_0}) \sim \frac{1}{t^{1/2}}$. Chandrasekhar's solution is valid for a planar weak shock pulse and simple wave flow behind the shock is assumed. The solution indicates that the distribution behind the shock is linear, that is, a triangular shock pulse. This Chandrasekhar's solution is for the decay of a weak triangular planar shock pulse.

2.4 Oswatitsch's Solution

A different approach was used by Oswatitsch (1956) to determine the decay of a weak "saw tooth" shock pulse. His method can be applied to cylindrical and spherical geometry provided that the shock profile is linear. The assumption of a "saw tooth" shock profile is not too restrictive because the far field shock profile in general, is indeed triangular for all geometries. Assuming isentropic flow behind a weak shock, the basic equations are given as,

$$\frac{2}{\gamma - 1}\left(\frac{\partial c}{\partial t} + u\frac{\partial c}{\partial r}\right) + c\frac{\partial u}{\partial r} + \frac{jcu}{r} = 0, \tag{2.4.1}$$

$$\frac{\partial u}{\partial t} + u\frac{\partial u}{\partial r} + \frac{2}{\gamma - 1}c\frac{\partial c}{\partial r} = 0, \tag{2.4.2}$$

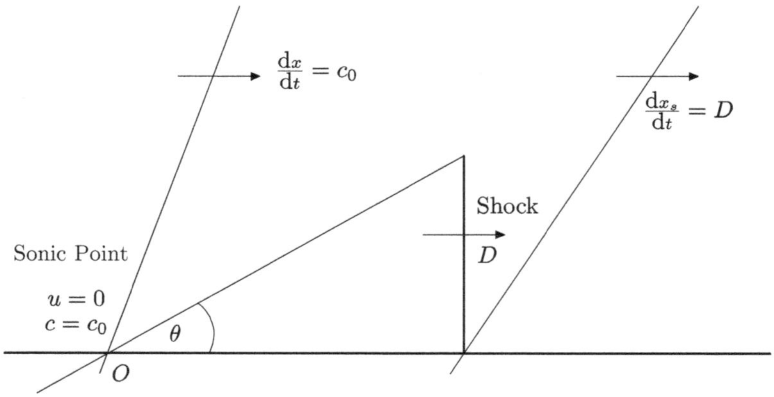

Figure 2.3 Linear Velocity Profile Behind a Shock Wave Traveling at Velocity D

where $j = 0, 1, 2$ for planar, cylindrical, and spherical geometries, respectively. Assuming a linear velocity profile, as shown in Fig. 2.3, there exists a sonic point O where $u = 0$, $c = c_0$ at the tail of the expansion fan behind the shock.

In the neighborhood of the sonic point O, Oswatitsch sought solutions of the form

$$\left.\begin{aligned} u(r,t) &= a_1(t)\varepsilon + a_2(t)\varepsilon^2 + \cdots \\ c(r,t) - c_0 &= b_1(t)\varepsilon + b_2(t)\varepsilon^2 + \cdots \end{aligned}\right\} \tag{2.4.3}$$

where $\varepsilon = r - c_0 t$ and $\varepsilon \ll 1$ near the sonic point O. Substituting the perturbation expressions (Eq 2.4.3) into the basic equations (Eqs 2.4.1 and 2.4.2) and collecting terms of the same order in ε, we obtain for zeroth order

$$a_1(t) = \frac{2}{\gamma - 1} b_1(t), \tag{2.4.4}$$

and for first order

$$\begin{aligned} \frac{2}{\gamma - 1}(\dot{b}_1 - 2b_2 c_0) + \frac{\gamma + 1}{\gamma - 1} a_1 b_1 + 2c_0 a_1 + \frac{j a_1}{t} &= 0, \\ \dot{a}_1 - 2c_0 a_2 + a_1^2 + \frac{2}{\gamma - 1}(2c_0 b_2 + b_1^2) &= 0, \end{aligned} \tag{2.4.5}$$

where $\dot{a}_1$ and $\dot{b}_1$ denote the differentiation with respect to time t. Eliminating a_2 and b_2 and using Eq 2.4.4, an equation for a_1 can be obtained as,

$$\dot{a}_1 + \frac{\gamma + 1}{2} a_1^2 + \frac{j a_1}{t} = 0. \tag{2.4.6}$$

Defining $\chi = \frac{1}{a_1 t}$, the above equation can be written as,

$$\dot{\chi} + \left(1 - \frac{j}{2}\right)\frac{\chi}{t} - \left(\frac{\gamma + 1}{2}\right)\frac{1}{t} = 0, \tag{2.4.7}$$

which can be integrated to yield

$$\chi(t) = \chi^\circ \left(\frac{t^\circ}{t}\right)^{1-j/2} + \left(\frac{\gamma+1}{2-j}\right)\left(1 - \left(\frac{t^\circ}{t}\right)^{1-j/2}\right). \tag{2.4.8}$$

for the planar case $j = 0$, and for the cylindrical case, $j = 1$. For the spherical case of $j = 2$, the solution for $\chi(t)$ is

$$\chi(t) = \chi^\circ + \ln\left(\frac{t}{t^\circ}\right)^{\frac{\gamma+1}{2}}. \tag{2.4.9}$$

In Eqs 2.4.8 and 2.4.9, the initial condition $t = t^\circ$, $\chi = \chi^\circ$ has been used. The solution for $\chi(t)$ gives the solution for $a_1(t)$ which represents the slope of the velocity profile at the sonic point. From Eq 2.4.3

$$a_1 = \left(\frac{\partial u}{\partial r}\right)_{\varepsilon=0}, \qquad b_1 = \left(\frac{\partial c}{\partial r}\right)_{\varepsilon=0} = \frac{\gamma-1}{2} a_1. \tag{2.4.10}$$

For linear distribution of u and c, the slopes a_1 and b_1 can be related to the pulse width $w(t)$ and the values u_1 and c_1 at the shock front, that is,

$$u_1 = \left(\frac{\partial u}{\partial r}\right)_{\varepsilon=0} w = a_1 w,$$

$$(c - c_0) = \left(\frac{\partial c}{\partial r}\right)_{\varepsilon=0} w = b_1 w = \left(\frac{\gamma-1}{2}\right) a_1 w.$$

From kinematics, $\frac{dw}{dt} = D - c_0$ and using Pfriem's formula, that is, $D = \frac{1}{2}(u_1 + c_1 + c_0)$, we can write

$$\frac{dw}{dt} = D - c_0 = \left[\frac{1}{2}(u_1 + c_1 + c_0)\right] - c_0 = \frac{1}{2}(u_1 + c_1 - c_0),$$

$$= \frac{1}{2}\left(a_1 w + \frac{\gamma-1}{2} a_1 w\right) = \frac{\gamma+1}{4} a_1 w.$$

Thus

$$\frac{1}{w}\frac{dw}{dt} = \frac{\gamma+1}{4} a_1 = \frac{\gamma+1}{4}\left(\frac{1}{\chi t}\right). \tag{2.4.11}$$

Using the solutions for $\chi(t)$, the pulse width $w(t)$ can be obtained from Eq 2.4.11 as

$$(j = 0), \qquad \left(\frac{w}{w_0}\right)^2 = 1 - \frac{\gamma+1}{2\chi^\circ}\left(1 - \frac{t}{t^\circ}\right), \tag{2.4.12}$$

$$(j = 1), \qquad \left(\frac{w}{w_0}\right)^2 = 1 - \frac{\gamma+1}{\chi^\circ}\left(1 - \left(\frac{t}{t^\circ}\right)^{\frac{1}{2}}\right), \tag{2.4.13}$$

$$(j = 2), \qquad \left(\frac{w}{w_0}\right)^2 = 1 + \frac{\gamma+1}{2\chi^\circ}\ln\left(\frac{t}{t^\circ}\right), \tag{2.4.14}$$

where $w = w^\circ$ at $t = t^\circ$. Since $\dfrac{dw}{dt} = D - c_0 = c_0(M_s - 1)$, we may express w as

$$w = \frac{4c_0(M_s - 1)}{(\gamma + 1)}(\chi t),$$

or

$$\chi = \frac{(\gamma + 1)w}{4c_0(M_s - 1)t}, \tag{2.4.15}$$

Defining χ° as

$$\chi^\circ = \frac{(\gamma + 1)w^\circ}{4c_0(M_s^\circ - 1)t^\circ}. \tag{2.4.16}$$

and replacing χ° in Eqs 2.4.12 to 2.4.14 by the above expression, we can write the solution for the pulse width "w" as

$$j = 0, \qquad \left(\frac{w}{w^\circ}\right)^2 = 1 - \frac{2(M_s^\circ - 1)}{w^\circ} c_0 t^\circ \left(1 - \frac{t}{t^\circ}\right), \tag{2.4.17}$$

$$j = 1, \qquad \left(\frac{w}{w^\circ}\right)^2 = 1 - \frac{4(M_s^\circ - 1)}{w^\circ} c_0 t^\circ \left(1 - \left(\frac{t}{t^\circ}\right)^{\frac{1}{2}}\right), \tag{2.4.18}$$

$$j = 2, \qquad \left(\frac{w}{w^\circ}\right)^2 = 1 + \frac{2(M_s^\circ - 1)}{w^\circ} c_0 t^\circ \ln\left(\frac{t}{t^\circ}\right), \tag{2.4.19}$$

where $M_s = M_s^\circ$ at $t = t^\circ$. Since $\dot{w} = c_0(M_s - 1)$, we may differentiate Eqs 2.4.17 to 2.4.19 and obtain

$$j = 0, \ (M_s - 1) = (M_s^\circ - 1)\left\{1 - \frac{2(M_s^\circ - 1)}{w^\circ} c_0 t^\circ \left(1 - \frac{t}{t^\circ}\right)\right\}^{-\frac{1}{2}}, \tag{2.4.20}$$

$$j = 1, \ (M_s - 1) = (M_s^\circ - 1)\left(\frac{t^\circ}{t}\right)^{\frac{1}{2}}\left\{1 - \frac{4(M_s^\circ - 1)}{w^\circ} c_0 t^\circ \left(1 - \left(\frac{t}{t^\circ}\right)^{\frac{1}{2}}\right)\right\}^{-\frac{1}{2}}, \tag{2.4.21}$$

$$j = 2, \ (M_s - 1) = (M_s^\circ - 1)\left(\frac{t^\circ}{t}\right)\left\{1 + \frac{2(M_s^\circ - 1)}{w^\circ} c_0 t^\circ \ln\left(\frac{t}{t^\circ}\right)\right\}^{-\frac{1}{2}}. \tag{2.4.22}$$

For large times, we can see that for planar waves $j = 0$, $M_s - 1 \sim \frac{1}{t^{1/2}}$. For cylindrical shocks, the asymptotic decay is $M_s - 1 \sim \frac{1}{t^{3/4}}$ and similarly for spherical waves, $M_s - 1 \sim \frac{1}{t \ln(t)^{1/2}}$. Since the shock overpressure $\Delta(\frac{p_1}{p_0}) = \frac{4\gamma}{\gamma+1}(M_s - 1)$, the decay of the shock overpressure behaves similarly.

Oswatitsch solution is valid for cylindrical and spherical waves but requires the profile to be linear.

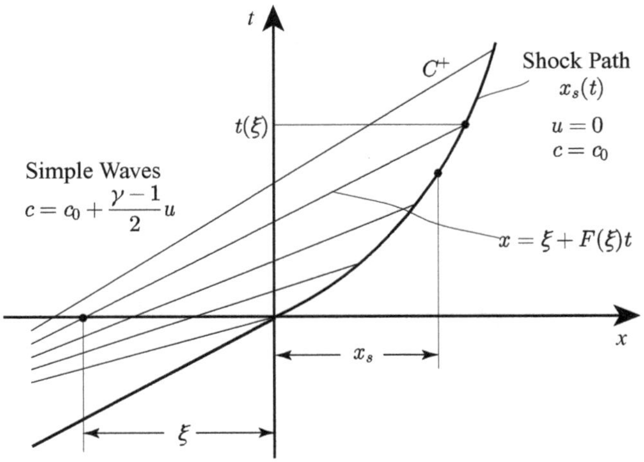

Figure 2.4 Expansion Waves Intersecting with Shock

2.5 Friedrichs' Theory

Friedrichs (1948) investigated various problems of formation and decay of weak planar shock waves. We shall consider only the case of the decay of a planar shock due to a non-steady expansion flow behind it. The shock is assumed to be uniform in the region $x < 0$, $t < 0$ and decays due to expansion waves behind it in the region $x > 0$, $t > 0$, and is shown in Fig. 2.4.

The flow behind the shock is assumed to be a simple wave flow where $c(u)$ is given by

$$c(u) = c_0 + \frac{\gamma - 1}{2} u, \tag{2.5.1}$$

and the trajectory of a C^+ is

$$x = \xi + F(\xi)t, \tag{2.5.2}$$

where

$$F(\xi) = c_0 + \frac{\gamma + 1}{2} u,$$

and $x = \xi$ at $t = 0$ is the origin of the C^+. Given the initial distribution of $u(\xi)$ (or $c(\xi)$), the trajectories of the C^+ are specified and, along each C^+, the values of u (and c) are constant, and correspond to their initial values. Thus the simple wave flow behind the shock is known. If the effects of the reflected characteristics on the flow can be neglected, then the simple wave flow behind the shock is independent of the shock motion. The decay of the shock can then be obtained by fitting the shock to the known simple wave flow specified initially. Different characteristics intersect the shock at different times, that is, the time of intersection $t(\xi)$ is a function of ξ. Hence the shock

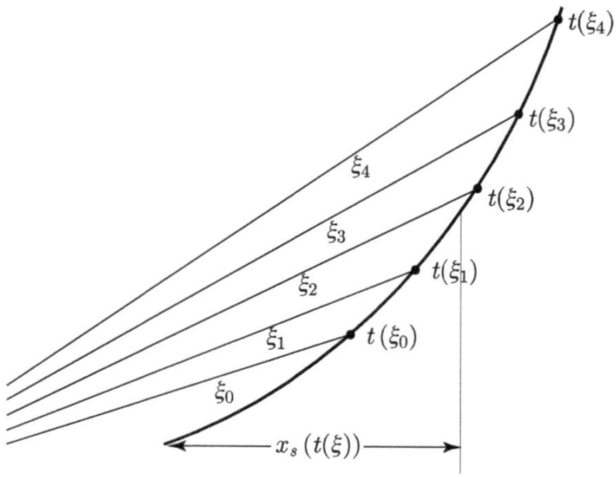

Figure 2.5 Characteristics Catching Up with the Shock

position x_s is a function of ξ, the characteristic that intersects it. This is illustrated in Fig. 2.5.

At time $t(\xi)$ of the intersection, the coordinates x_s, $t(\xi)$ correspond to a point on the trajectory of the ξ characteristic. Thus

$$x = x_s = \xi + F(\xi)t(\xi). \tag{2.5.3}$$

The above equation simply denotes that at a particular time $t(\xi)$ the position x on the trajectory of the ξ characteristic coincides with the shock position x_s. We wish to obtain the dependence of $t(\xi)$ on ξ. To obtain the equation that gives $t(\xi)$, we differentiate Eq 2.5.3 with respect to ξ along the shock path, that is,

$$\frac{\mathrm{d}x_s\,(t(\xi))}{\mathrm{d}\xi} = \frac{\mathrm{d}x_s}{\mathrm{d}t}\frac{\mathrm{d}t(\xi)}{\mathrm{d}\xi} = \dot{x}_s\frac{\mathrm{d}t(\xi)}{\mathrm{d}\xi} = 1 + t(\xi)\frac{\mathrm{d}F(\xi)}{\mathrm{d}\xi} + F(\xi)\frac{\mathrm{d}t(\xi)}{\mathrm{d}\xi}.$$

And re-arranging, the above equation can be written as

$$(F(\xi) - \dot{x}_s)\frac{\mathrm{d}t(\xi)}{\mathrm{d}\xi} + 1 + t(\xi)\frac{\mathrm{d}F(\xi)}{\mathrm{d}\xi} = 0. \tag{2.5.4}$$

Solution of the above equation gives $t(\xi)$ as a function of ξ. When the characteristic intersects the shock, we require the shock strength to be such that the state behind the shock (as given by the Rankine–Hugoniot equations), matches the state carried by the characteristic that intersects it. If we consider matching the particle velocity, the Rankine–Hugoniot gives

$$\frac{u_1}{c_0} = \frac{2}{\gamma+1}\left(M_s - \frac{1}{M_s}\right) = \frac{2}{\gamma+1}\frac{(M_s+1)(M_s-1)}{M_s}, \tag{2.5.5}$$

and for $M_s \simeq 1$, $M_s + 1 \simeq 2$, and defining $\sigma_1 = \dfrac{\gamma+1}{2}\dfrac{u_1}{c_0}$, Eq 2.5.5 can be written as

$$M_s = \frac{\dot{x}_s}{c_0} = \left(1 - \frac{\sigma_1}{2}\right)^{-1} = 1 + \frac{\sigma_1}{2} + \frac{\sigma_1^2}{8} + \cdots \tag{2.5.6}$$

The particle velocity corresponding to the characteristic can be found from the slope

$$F(\xi) = u + c = c_0 + \frac{\gamma + 1}{2} u(\xi) = c_0(1 + \sigma(\xi)). \tag{2.5.7}$$

Matching the particle velocity, we equate $\sigma_1 = \sigma$, and substituting Eqs 2.5.6 and 2.5.7 into Eq 2.5.4 and rearranging yields the equation for $t(\xi)$ as

$$\frac{\mathrm{d}(c_0 t)}{\mathrm{d}\xi} + \frac{8(c_0 t)}{(4 - \sigma)\sigma} \frac{\mathrm{d}\sigma}{\mathrm{d}\xi} = -\frac{8}{\sigma(4 - \sigma)}. \tag{2.5.8}$$

The above equation can be integrated to yield

$$c_0 t = 4A \left(\frac{1}{\sigma} - \frac{1}{4} \right)^2, \tag{2.5.9}$$

where

$$A = 32 \int_{\xi}^{0} \frac{\sigma(\xi')\mathrm{d}\xi'}{[4 - \sigma(\xi')]^3}. \tag{2.5.10}$$

In Eq 2.5.10, ξ' is a dummy integration variable. Given the initial distribution $u(\xi)$, (i.e., $\sigma(\xi)$), the integral A can be evaluated. Equation 2.5.9 gives the solution for $t(\xi)$ in terms of $\sigma(\xi)$. Substituting Eq 2.5.9 into Eq 2.5.3, we write the shock trajectory in terms of $\sigma(\xi)$ as

$$x_s = \xi + A(1 + \sigma) \left(\frac{4 - \sigma}{2\sigma} \right)^2. \tag{2.5.11}$$

Alternately, solving Eq 2.5.9 for $\sigma(t)$ gives

$$\sigma(t) = 2\sqrt{\frac{A}{c_0 t}} \left(1 + \frac{1}{2}\sqrt{\frac{A}{c_0 t}} \right)^{-1}. \tag{2.5.12}$$

Substituting the above for $\sigma(t)$ in Eq 2.5.3 gives the shock path in terms of t as

$$x_s(t) = \xi + c_0 t + 2\sqrt{A c_0 t} \left(1 + \frac{1}{2}\sqrt{\frac{A}{c_0 t}} \right)^{-1}. \tag{2.5.13}$$

With $\sigma(t)$ given by Eq 2.5.12 the other shock properties can be obtained from the Rankine–Hugoniot equations, that is,

$$\left. \begin{aligned} \frac{u}{c_0} &= \frac{2}{\gamma + 1}\sigma, & \frac{c}{c_0} &= 1 + \left(\frac{\gamma - 1}{\gamma + 1} \right)\sigma \\ \frac{\Delta p}{p_0} &= \frac{2\gamma}{\gamma + 1}\sigma, & \frac{\rho}{\rho_0} &= 1 + \frac{2}{\gamma + 1}\sigma \end{aligned} \right\} \tag{2.5.14}$$

For large times where $t \to \infty$, $M_s \to 1$, $\sigma \to 0$, we may expand Eq 2.5.12 and write

$$\sigma(t) = 2\sqrt{\frac{A}{c_0 t}} \left(1 - \frac{1}{2}\sqrt{\frac{A}{c_0 t}} + \cdots \right) = 2\sqrt{\frac{A}{c_0 t}} - \frac{A}{c_0 t} \approx 2\sqrt{\frac{A}{c_0 t}}. \tag{2.5.15}$$

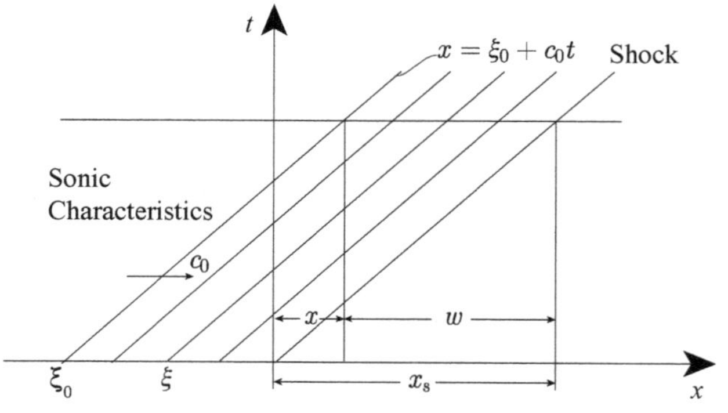

Figure 2.6 Width of the Shock Pulse with Time

Since the shock overpressure $\frac{\Delta p_1}{p_0} = \frac{2\gamma}{\gamma+1}\sigma$, the asymptotic decay for the shock overpressure is thus $\frac{\Delta p_1}{p_0} \sim \frac{1}{t^{1/2}}$, which is in accord with Chandrasekhar's solution

All the characteristics $\xi_0 \le \xi \le 0$ will eventually catch up with the shock, but for large time $t \to \infty$, the shock path will asymptotically be parallel to the sonic characteristic ξ_0 given by $x = \xi_0 + F(\xi)t = \xi_0 + c_0 t$. The width of the shock pulse', shown in Fig. 2.6, will be given by $w = x_s - (\xi_0 + c_0 t)$ and using Eq 2.5.15, we write

$$w = x_s - (\xi_0 + c_0 t) = 2\sqrt{Ac_0 t} - A + \cdots \tag{2.5.16}$$

Thus the pulse width increase like $w \sim t^{\frac{1}{2}}$ in contrast to the acoustic solution where $w =$ constant.

The asymptotic profiles for u and c can be obtained from σ. Writing σ in the neighborhood of ξ_0 as

$$\sigma = a(\xi - \xi_0) + b(\xi - \xi_0)^2 + \cdots \tag{2.5.17}$$

and substituting the above into Eq 2.5.1, we obtain

$$\begin{aligned} x &= \xi + c_0 t(1 + \sigma), \\ &= [\xi_0 + (\xi - \xi_0)] + c_0 t\left[1 + a(\xi - \xi_0) + b(\xi - \xi_0)^2 + \cdots\right], \\ &= (\xi_0 + c_0 t) + (\xi - \xi_0) + ac_0 t(\xi - \xi_0) + bc_0 t(\xi - \xi_0)^2 + \cdots \end{aligned}$$

which can be re-arranged to read

$$\xi - \xi_0 = \frac{x - (\xi_0 + c_0 t)}{1 + ac_0 t}\left(1 + \frac{bc_0 t(\xi - \xi_0)}{1 + ac_0 t}\right)^{-1}.$$

Expanding the last term, we write

$$\xi - \xi_0 = \frac{x - (\xi_0 + c_0 t)}{1 + ac_0 t}\left(1 - \frac{bc_0 t(\xi - \xi_0)}{1 + ac_0 t} + \cdots\right),$$

and replacing $(\xi - \xi_0)$ in the term of the above by $\frac{x-(\xi_0+c_0t)}{1+ac_0t}$, we can write

$$\xi - \xi_0 \simeq \frac{x-(\xi_0+c_0t)}{1+ac_0t}\left(1 - \frac{bc_0t(\xi-\xi_0)}{(1+ac_0t)^2}\right),$$

hence

$$(\xi-\xi_0)^2 = \frac{[x-(\xi_0+c_0t)]^2}{(1+ac_0t)^2} - \frac{2bc_0t\,[x-(\xi_0+c_0t)]^3}{(1+ac_0t)^4}.$$

Substituting the above into Eq 2.5.17 yields

$$\sigma(\xi) = a\left(\frac{x-(\xi_0+c_0t)}{1+ac_0t}\right) + \frac{b\,(x-(\xi_0+c_0t))^2}{(1+ac_0t)^3} + \cdots$$

Thus, we note that up to first order in $x-(\xi_0+c_0t)$, σ is linear with x at any given time t. Accordingly, the asymptotic distributions of u and c in the shock pulse $\xi_0+c_0t \le x \le \xi_0+c_0t+w(t)$ are also linear.

2.6 Decay of a Piston Driven Shock

As an example to illustrate Freidrichs' theory, we shall consider the decay of a weak planar shock. At $t=0$, a piston moving at a constant velocity u_1 drives a shock ahead of it. At $t=t_R$, the piston is stopped abruptly, and a center rarefaction fan is generated. The shock starts to decay at $t=t_1$ when the head of the expansion fan catches up to the shock front. We shall consider the shock to be weak.

The wave processes are illustrated in Fig. 2.7.

Neglecting changes in the Riemann invariant across the weak shock, $J^- = u - \frac{2}{\gamma-1}c = \text{constant}$. The flow behind the shock is a simple wave and

$$c = c_0 + \frac{\gamma-1}{2}u. \tag{2.6.1}$$

Along a C^+, $J^+ = u + \frac{2}{\gamma-1}c = \text{constant}$ and from Eq 2.6.1, we write

$$J^+ = u + \frac{2}{\gamma-1}c = \frac{2}{\gamma-1}\left(c + \frac{\gamma-1}{2}u\right) = \frac{2}{\gamma-1}(2c-c_0).$$

Thus each C^+ in the simple wave region is characterized by a value of u (or c). The equation for a C^+ is given by

$$\frac{dx}{dt} = u + c = c_0 + \frac{\gamma+1}{2}u = c_0(1+\sigma),$$

where $\sigma = \frac{\gamma+1}{2}\frac{u}{c_0}$ and was defined in Eq 2.5.6. Integrating yields the trajectory of a C^+ as

$$x = \left(c_0 + \frac{\gamma+1}{2}u\right)t + f(u) = c_0(1+\sigma)t + f(u). \tag{2.6.2}$$

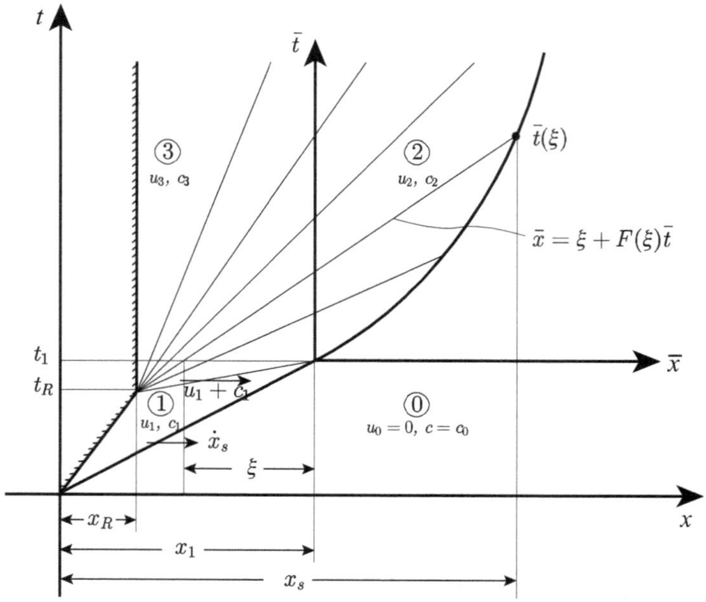

Figure 2.7 Wave Processes in the Decay of a Piston Driven Shock

where the integration constant $f(u)$ can be determined from initial or boundary conditions.

From kinematics, we can write

$$x_R = u_1 t_R = \frac{2c_0}{\gamma + 1}\sigma_1 t_R,$$

$$x_1 = x_R + (u_1 + c_1)t_1 = \dot{x}_s t_1 = x_R + c_0(1 + \sigma_1)t_1, \tag{2.6.3}$$

since $x_1 = \dot{x}_s t_1$. For a weak shock we can write

$$\dot{x}_s = c_0\left(1 + \frac{\sigma_1}{2} + \frac{\sigma_1^2}{8}\right),$$

and solving for t_1, we obtain

$$t_1 = \frac{8\left(1 + \frac{\gamma-1}{\gamma+1}\sigma_1\right)t_R}{\sigma_1(4 - \sigma_1)}. \tag{2.6.4}$$

Given u_1 (hence σ_1) and t_R, x_p, and t_1 can be determined.

The solution for the centered rarefaction fan is given by

$$x - x_R = (u + c)(t - t_R) = c_0(1 + \sigma)(t - t_R),$$

which can be re-arranged to read

$$x - x_1 = (x_R - x_1) + c_0(1 + \sigma)(t_1 - t_R) + c_0(1 + \sigma)(t - t_1).$$

The above equation can be written as

$$\bar{x} = \xi + c_0(1+\sigma)\bar{t}, \tag{2.6.5}$$

where

$$\begin{aligned} \xi &= (x_R - x_1) + c_0(1+\sigma)(t_1 - t_R), \\ &= c_0(\sigma - \sigma_1)(t_1 - t_R), \end{aligned} \tag{2.6.6}$$

and $\bar{x} = x - x_1$ and $\bar{t} = t - t_1$. To determine the decay of the shock in the region $\bar{x} > 0$, $\bar{t} > 0$, we solve for $\bar{t}(\xi)$, that is,

$$c_0\bar{t} = 8\left(\frac{4-\sigma}{\sigma}\right)^2 \int_\xi^0 \frac{\sigma(\xi')\mathrm{d}\xi'}{[4-\sigma(\xi')]^3} = 4A\left(\frac{1}{\sigma} - \frac{1}{4}\right)^2, \tag{2.6.7}$$

where

$$A = 32\int_\xi^0 \frac{\sigma(\xi')\mathrm{d}\xi'}{[4-\sigma(\xi')]^3}.$$

From Eq 2.6.6, we obtain

$$\mathrm{d}\xi' = c_0(t_1 - t_R)\mathrm{d}\sigma'.$$

Thus the integral A can be written as

$$A = 32\int_0^{\sigma_1} \frac{\sigma'\mathrm{d}\sigma'}{[4-\sigma(\xi')]^3},$$

where σ' is a dummy integration variable. Evaluating the integral gives

$$A = c_0(t_1 - t_R)\left[\left(\frac{\sigma_1}{4-\sigma_1}\right)^2 - \left(\frac{\sigma}{4-\sigma}\right)^2\right], \tag{2.6.8}$$

and substituting the above into Eq 2.6.7 yields

$$\bar{t} = (t_1 - t_R)\left[\left(\frac{4-\sigma}{\sigma}\right)^2\left(\frac{\sigma_1}{4-\sigma_1}\right)^2 - 1\right],$$

and re-arranging, we can write

$$t = t_R + (t_1 - t_R)\left[\left(\frac{4-\sigma}{\sigma}\right)^2\left(\frac{\sigma_1}{4-\sigma_1}\right)^2\right]. \tag{2.6.9}$$

Solving for σ in Eq 2.6.7, and using the above equation, we get

$$\sigma(t) = \frac{4}{1 + \left(\frac{4-\sigma_1}{\sigma_1}\right)\sqrt{\frac{t-t_R}{t_1-t_R}}},$$

or alternately, we write

$$\sigma(t) = 4\left(\frac{\sigma_1}{4-\sigma_1}\right)\sqrt{\frac{t_1-t_R}{t-t_R}}\left(1 + \left(\frac{\sigma_1}{4-\sigma_1}\right)\sqrt{\frac{t_1-t_R}{t-t_R}}\right)^{-1}. \tag{2.6.10}$$

The shock trajectory can now be written as

$$x_s = x_R + c_0(1+\sigma)(t-t_R),$$

$$= x_R + c_0\left[(t-t_R) + \frac{(t-t_R)}{1+\left(\frac{4-\sigma_1}{\sigma_1}\right)\sqrt{\frac{t-t_R}{t_1-t_R}}}\right].$$

Alternately, we can write the shock trajectory as

$$x_s = x_R + c_0\left[(t-t_R) + \frac{4\sqrt{(t_1-t_R)(t-t_R)}}{K\left(1+\frac{1}{K}\sqrt{\frac{t_1-t_R}{t-t_R}}\right)}\right], \tag{2.6.11}$$

where $K = \frac{4-\sigma_1}{\sigma}$. Large times $\sigma(t)$ and the shock path x_s can be written as

$$\sigma(t) = 4K\sqrt{\frac{t_1-t_R}{t-t_R}} - 4K^2\left(\frac{t_1-t_R}{t-t_R}\right) + \cdots \tag{2.6.12}$$

$$x_s = x_R + c_0(t-t_R) + \frac{4c_0}{K}\sqrt{(t_1-t_R)(t-t_R)} - \frac{4c_0}{K^2}(t_1-t_R). \tag{2.6.13}$$

Since $t \gg t_R$, and defining

$$A = \frac{4c_0}{K^2}(t_1-t_R),$$

the shock path can be written as

$$x_s = x_R + c_0(t-t_R) + 2\sqrt{Ac_0(t-t_R)} - A$$

$$= \xi_0 + c_0 t + 2\sqrt{Ac_0 t} - A, \tag{2.6.14}$$

where $\xi_0 = x_R - c_0 t_R$ and we have neglected t_R when compared to t for large times. Thus for large times $\sigma \sim \frac{1}{t^{1/2}}$, thus $\frac{\Delta p_1}{p_0} \sim \frac{1}{t^{1/2}}$ for the shock overpressure decay. The width of the shock pulse $w(t) = x_s - (\xi_0 + c_0 t)$, since $\xi_0 + c_0 t$ is the trajectory of the sonic characteristic $\frac{dx}{dt} = c_0$. We also note from Eq 2.6.14 that for large times, w increases like $t^{\frac{1}{2}}$, that is,

$$w(t) \simeq 2\sqrt{Ac_0 t} - A \sim t^{\frac{1}{2}}.$$

This result corresponds the solution of Chandrasekhar described previously.

2.7 Whitham's Theory

Whitham (1956) developed a method to describe the far-field decay of a weak shock wave. His method is very similar to Friedrichs' theory and differs in that acoustic theory is used to describe the flow field behind the shock instead of using finite amplitude simple waves. Linear acoustic theory gives a wave of permanent form and fails to describe the non-linear distortion in the far-field when the cumulative effect of the small differences in the speed of the characteristics becomes significant. To describe the decay of the shock, Whitham "non-linearized" the acoustic solution by replacing the acoustic trajectories by the local sound speed, that is, characteristics. Whitham's so-called modified acoustic theory can therefore describe the distortion of the wave form as the characteristics are now no longer parallel.

With geometric acoustic theory in mind, Whitham considered the wave propagation in a ray tube of cross-sectional area $A(s)$, where s is the distance along the ray tube. Acoustic theory for the solution of the velocity potential can be written as

$$\phi = \frac{f(t - \frac{s}{c_0})}{A(s)^{\frac{1}{2}}},$$

and the particle velocity can be obtained as

$$u = \frac{\partial \phi}{\partial s} = \frac{-f'\left(t - \frac{s}{c_0}\right)}{c_0 A(s)^{\frac{1}{2}}} - \frac{-f\left(t - \frac{s}{c_0}\right)}{2A(s)^{\frac{3}{2}}}\frac{\mathrm{d}A(s)}{\mathrm{d}s},$$
$$\simeq \frac{-f'\left(t - \frac{s}{c_0}\right)}{c_0 A(s)^{\frac{1}{2}}}.$$

The second term can be neglected since it is small as compared to the first term because $A(s)$ is, in general, an increasing function of s. The pressure perturbation is given as

$$\Delta p = -\rho_0 \frac{\partial \phi}{\partial t} = \frac{f'\left(t - \frac{s}{c_0}\right)}{A(s)^{\frac{1}{2}}} = \gamma p_0 \frac{u}{c_0}.$$

For an isentropic process,

$$\frac{c}{c_0} = \left(\frac{p}{p_0}\right)^{\frac{\gamma-1}{2\gamma}},$$

or

$$\left(\frac{\Delta c}{c_0} + 1\right) = \left(1 + \frac{\Delta p}{p_0}\right)^{\frac{\gamma-1}{2\gamma}},$$

giving $\dfrac{\Delta c}{c_0} = \dfrac{\gamma - 1}{2\gamma}\dfrac{\Delta p}{p_0} + \cdots$. In Whitham's modified acoustic theory, the argument of the function f corresponds to the acoustic trajectory $\tau = t - \frac{s}{c_0}$ in the near field. For large s, in the far field, the characteristic $\tau = u + c$ is used. The modified acoustic solution can be written as

$$\frac{\Delta p}{p_0} = \frac{G(\tau)}{A(s)^{\frac{1}{2}}}, \tag{2.7.1}$$

where

$$G(\tau) = -\frac{\gamma}{c_0^2} f'(\tau), \tag{2.7.2}$$

and accordingly, the particle velocity and sound speed are given as

$$\frac{u}{c_0} = -\frac{f'(\tau)}{c_0^2 A(s)^{\frac{1}{2}}} = \frac{G(\tau)}{\gamma A(s)^{\frac{1}{2}}} = \frac{1}{\gamma}\frac{\Delta p}{p_0}, \tag{2.7.3}$$

$$\frac{\Delta c}{c_0} = \frac{\gamma - 1}{2\gamma}\frac{\Delta p}{p_0} = \frac{\gamma - 1}{2\gamma}\frac{G(\tau)}{\gamma A(s)^{\frac{1}{2}}}. \tag{2.7.4}$$

The equation for a characteristic τ = constant can be written as

$$\left.\frac{\mathrm{d}s}{\mathrm{d}t}\right)_{\tau=\text{constant}} = u + c = c_0\left(\frac{u}{c_0} + 1 + \frac{\Delta c}{c_0}\right) = c_0\left(1 + \frac{\gamma+1}{2\gamma}\frac{\Delta p}{p_0}\right)$$

$$= \frac{1}{c_0}\left(1 + \frac{\gamma+1}{2\gamma}\frac{\Delta p}{p_0}\right)^{-1} = \frac{1}{c_0}\left(1 - \frac{\gamma+1}{2\gamma}\frac{\Delta p}{p_0} + \cdots\right)$$

$$= \frac{1}{c_0}\left(1 - \frac{\gamma+1}{2\gamma}\frac{G(\tau)}{A(s)^{\frac{1}{2}}} + \cdots\right).$$

Integrating the above gives the trajectory of the characteristic τ = constant, that is,

$$t = \frac{s}{c_0} - KG(\tau)\int_0^s \frac{\mathrm{d}s}{A(s)^{1/2}} + f(\tau),$$

where $f(\tau)$ is the constant of integration. We can simply take $f(\tau) = \tau$ itself so that in the near field when s is small, $\tau = t - \frac{s}{c_0}$ corresponding to the acoustic trajectory since the integral term is small. Thus we write

$$t = \frac{s}{c_0} + \tau - KG(\tau)\int_0^s \frac{\mathrm{d}s}{A(s)^{\frac{1}{2}}}, \tag{2.7.5}$$

where $K = \frac{\gamma+1}{2\gamma c_0}$. The various terms in Eq 2.7.5 are illustrated in Fig. 2.8. Different characteristic intersects the shock when it is at different position s. Characteristics behind the shock continuously catch up with the shock causing it to decay, as shown in Fig. 2.9. Let $\tau = T(s)$ be the particular characteristic that intersects the shock when it is at position s. When the characteristic intersects the shock, the shock coordinates (s, t) will be a point on the particular characteristic $\tau = T(s)$. Thus we write Eq 2.7.5 as

$$t = \frac{s}{c_0} + T(s) - KG(\tau)\int_0^s \frac{\mathrm{d}s}{A(s)^{\frac{1}{2}}}. \tag{2.7.6}$$

Equation 2.7.6 gives the shock coordinates parametrically in terms of the characteristic $\tau = T(s)$ that intersects it. To obtain the dependence of $T(s)$ on shock positions, we differentiate Eq 2.7.6 and obtain

$$\frac{\mathrm{d}t}{\mathrm{d}s} = \frac{1}{D} = \frac{1}{c_0} + \frac{\mathrm{d}T(s)}{\mathrm{d}s} - \frac{KG(\tau)}{A(s)^{\frac{1}{2}}} - \frac{K\mathrm{d}G(\tau)}{\mathrm{d}T}\frac{\mathrm{d}T}{\mathrm{d}s}\int_0^s \frac{\mathrm{d}s}{A(s)^{\frac{1}{2}}}, \tag{2.7.7}$$

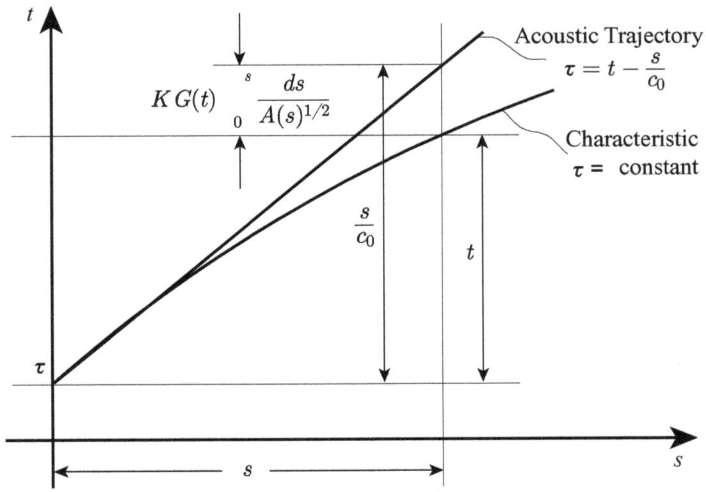

Figure 2.8 Terms in Eq 2.7.5

where $D = \frac{ds}{dt}$ is the shock speed. The Rankine–Hugoniot equations give the shock speed (or M_s) in terms of the overpressure, that is,

$$D = c_0 M_s = c_0 \left(1 + \frac{\gamma + 1}{4\gamma} \frac{\Delta p_1}{p_0}\right).$$

For weak shocks where $\frac{\Delta p_1}{p_0} \ll 1$, we write

$$\frac{1}{D} = \frac{1}{c_0}\left(1 + \frac{\gamma + 1}{4\gamma}\frac{\Delta p_1}{p_0}\right)^{-1} = \frac{1}{c_0}\left(1 - \frac{\gamma + 1}{4\gamma}\frac{\Delta p_1}{p_0} + \cdots\right).$$

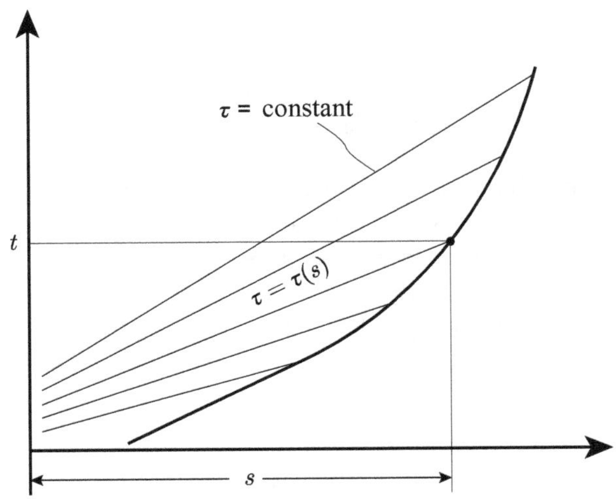

Figure 2.9 Decay of Shock by the Characteristics Catching Up with It

When the characteristic intersects the shock, the shock strength must be such that the shock overpressure corresponding to the characteristic value, that is, $\frac{\Delta p_1}{p_0} = \frac{\Delta p}{p_0} = \frac{G(\tau)}{A(s)^{\frac{1}{2}}}$. Thus

$$\frac{1}{D} = \frac{1}{c_0} - \frac{K}{2}\frac{G(\tau)}{A(s)^{\frac{1}{2}}},$$

and substituting the above into Eq 2.7.7 gives

$$\frac{\mathrm{d}T(s)}{\mathrm{d}s} = \frac{K}{2}\frac{G(\tau)}{A(s)^{\frac{1}{2}}} + \frac{K\mathrm{d}G(\tau)}{\mathrm{d}T}\frac{\mathrm{d}T(s)}{\mathrm{d}s}\int_0^s \frac{\mathrm{d}s}{A(s)^{\frac{1}{2}}},$$

which can be integrated if we multiply the above equation by $2G(\tau)$ and re-arranging, to give

$$\frac{\mathrm{d}}{\mathrm{d}s}\left[KG^2(\tau)\int_0^s \frac{\mathrm{d}s}{A(s)^{\frac{1}{2}}}\right] = 2G(\tau)\frac{\mathrm{d}T}{\mathrm{d}s},$$

which integrates to yield

$$G(\tau) = \left[\frac{\frac{2}{K}\int_0^T G(T')\mathrm{d}T'}{\int_0^s \frac{\mathrm{d}s}{A(s)^{\frac{1}{2}}}}\right]^{\frac{1}{2}}, \tag{2.7.8}$$

where T' is a dummy integration variable. We have assumed for definiteness that $T = 0$ at $s = 0$. Equation 2.7.8 gives the solution $G(T)$ corresponding to the particular characteristic $\tau = T(s)$ that intersects the shock at time t when it is at position s. From the solution $G(T)$, the shock overpressure, particle velocity, and sound speed can all be determined (i.e., Eqs 2.7.1, 2.7.3, 2.7.4). Substituting Eq 2.7.8 into 2.7.6 gives the shock path, that is,

$$t = \frac{s}{c_0} + T(s) - \left\{2K\int_0^T G(\xi)\mathrm{d}\xi \int_0^s \frac{\mathrm{d}s}{A(s)^{\frac{1}{2}}}\right\}^{\frac{1}{2}}, \tag{2.7.9}$$

where ξ is a dummy integration variable. Equation 2.7.9 gives the shock path parametrically in terms of $T(s)$, the characteristic that intersects it.

In the far field, the shock profile in general takes on the shape of an N-wave with a sonic point that propagates at c_0, where $u = 0$, $c = c_0$, $\frac{\Delta p}{p_0} = 0$ (at the sonic point). The sonic trajectory is given by $\tau_0 = t - \frac{s}{c_0}$ and the solution $G(\tau_0) = 0$. Since $A(s)$ is in general, an increasing function of s, $\int_0^s \frac{\mathrm{d}s}{A^{\frac{1}{2}}} \to \infty$ as $s \to 0$. The solution $G(\tau)$ for large values of s can be approximated by

$$G(T) = \left[\frac{\frac{2}{K}\int_0^{T_0} G(\xi)\mathrm{d}\xi}{\int_0^s \frac{\mathrm{d}s}{A(s)^{\frac{1}{2}}}}\right]^{\frac{1}{2}}, \tag{2.7.10}$$

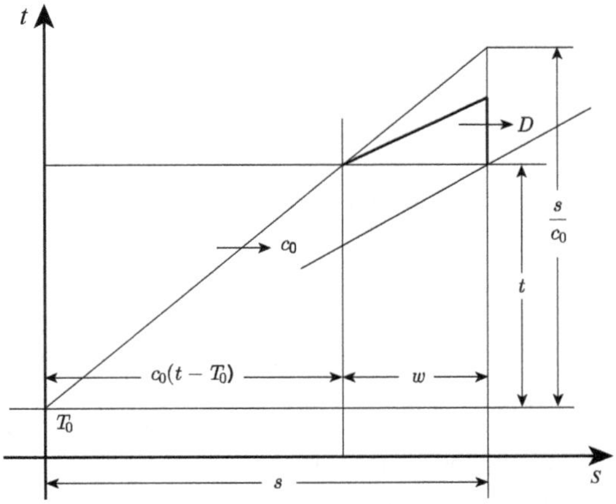

Figure 2.10 Shock Front Leading the Sonic Characteristic by Pulse Width

and thus the overpressure behind the shock as $s \to \infty$ can be written as

$$\frac{\Delta p}{p_0} = \frac{G(\tau)}{A(s)^{\frac{1}{2}}} = \left[\frac{\frac{2}{K}\int_0^{T_0} G(\xi)\mathrm{d}\xi}{A(s)\int_0^s \frac{\mathrm{d}s}{A(s)^{\frac{1}{2}}}} \right]^{\frac{1}{2}}. \tag{2.7.11}$$

For planar, cylindrical, and spherical geometries, $A(s) \sim$ constant, $A(s) \sim s$, and $A(s) \sim s^2$, respectively. Thus in the asymptotic limit where $s \to \infty$, $\frac{\Delta p}{p_0} \sim \frac{1}{s^{1/2}}$, $\frac{\Delta p}{p_0} \sim \frac{1}{s^{3/4}}$, and $\frac{\Delta p}{p_0} \sim \frac{1}{s \ln s^{\frac{1}{2}}}$ for the planar, cylindrical, and spherical geometries, respectively. This result is in accord with Oswatitsch's solution described in a previous section.

For large values of s the shock trajectory can be approximated by

$$t = \frac{s}{c_0} + T_0 - \left\{ 2K \int_0^{T_0} G(\xi)\mathrm{d}\xi \int_0^s \frac{\mathrm{d}s}{A(s)^{\frac{1}{2}}} \right\}^{\frac{1}{2}}. \tag{2.7.12}$$

Since the shock front is ahead of the sonic characteristic by the pulse width w as illustrated in Fig. 2.10, we write $s = c_0(t - T_0) + w$ or $\dfrac{w}{c_0} = \dfrac{s}{c_0} - t + T_0$. From Eq 2.7.12, we obtain

$$\frac{w}{c_0} = \left\{ 2K \int_0^{T_0} G(\xi)\mathrm{d}\xi \int_0^s \frac{\mathrm{d}s}{A(s)^{\frac{1}{2}}} \right\}^{\frac{1}{2}}, \tag{2.7.13}$$

in contrast to the linear acoustic theory which gives $w =$ constant.

For large values of s, we may write $G(\tau)$ in the neighborhood of the sonic characteristics using Eq 2.7.12 as

$$G(\tau) = -\frac{\left(t - \frac{s}{c_0} - T_0\right)}{K} \left[\int_0^s \frac{\mathrm{d}s}{A(s)^{\frac{1}{2}}} \right]^{-1},$$

Thus, the overpressure in the neighborhood of T_0 is

$$\frac{\Delta p}{p_0} = \frac{G(\tau)}{A(s)^{\frac{1}{2}}} = -\frac{\left(t - \frac{s}{c_0} - T_0\right)}{KA(s)^{\frac{1}{2}}}\left[\int_0^s \frac{\mathrm{d}s}{A(s)^{\frac{1}{2}}}\right]^{-1}.$$

At any fixed distance s the overpressure decays linearly with time and the rate of the pressure decrease is

$$\frac{\partial}{\partial t}\left(\frac{\Delta p}{p_0}\right) = -\frac{1}{K}\left[A^{\frac{1}{2}}\int_0^s \frac{\mathrm{d}s}{A(s)^{\frac{1}{2}}}\right]^{-1}.$$

The above indicates that the pressure decay is not dependent on the initial wave profile $G(\tau)$, and is a function of the shock position s only.

As an example of Whitham's theory, consider the far-field decay of the spherical shock generated by a constant velocity expanding piston. $r_p = \alpha c_0 t$ where $\alpha \ll 1$. The acoustic solution of this problem has been given by G. I. Taylor, that is,

$$f(r - c_0 t) = \frac{\alpha^3 c_0^3}{\alpha^2 - 1}(r - c_0 t)^2 = \frac{\alpha^3 c_0^3}{\alpha^2 - 1}\tau^2.$$

The solution $G(\tau)$ can be written as

$$G(\tau) = -\frac{r}{c_0^2} f' = -\frac{r}{c_0^2}\frac{2\alpha^3 c_0^3 \tau}{(\alpha^2 - 1)} \simeq 2\gamma\alpha^3 c_0 \tau,$$

since $\alpha \ll 1$. Further since $G(\tau)$ is continuous at $\tau = 0$, there is no leading shock front according to acoustic theory. However, according to the modified acoustic theory, the wave profile will be distorted leading to the formation of a shock wave in the far field. From Eq 2.7.10, we write

$$\int_{r_0}^{r} \frac{\mathrm{d}r}{A(s)^{\frac{1}{2}}} = \frac{2}{K}\frac{\int_0^T 2\gamma\alpha^3 c_0 \xi \,\mathrm{d}\xi}{(2\gamma\alpha^3 c_0 T)^2}.$$

Thus

$$\frac{\ln r}{r_p} = \frac{1}{K 2\gamma\alpha^3 c_0} = \frac{1}{(\gamma + 1)\alpha^3},$$

where $K = \dfrac{\gamma + 1}{2\gamma c_0}$.

Replacing r_p by $\alpha c_0 T$, we get

$$\ln\frac{r}{\alpha c_0 T} = \frac{1}{(\gamma + 1)\alpha^3},$$

and the overpressure can be obtained as

$$\frac{\Delta p}{p_0} = \frac{G(\tau)}{r} = \frac{2\gamma\alpha^3 c_0 T}{\alpha c_0 T \exp\frac{1}{(\gamma+1)\alpha^3}}.$$

Thus

$$\frac{\Delta p}{p_0} = 2\gamma\alpha^2 \exp\left(-\frac{1}{\alpha^3(\gamma+1)}\right),$$

and since $(M_s - 1) = \dfrac{\gamma+1}{4\gamma}\frac{\Delta p}{p_0}$, the shock Mach number can be obtained as

$$(M_s - 1) = \frac{\gamma+1}{2}\alpha^2 \exp\left(-\frac{1}{\alpha^3(\gamma+1)}\right).$$

Whitham pointed out that the accuracy of the pre-exponential term was uncertain and that error terms may well be more dominant. We may write

$$\ln(M_s - 1) \sim \frac{1}{\alpha^3(\gamma+1)},$$

which is in the form given by Lighthill (1948).

3 Shock Propagation in a Non-uniform Cross-sectional Area Tube

3.1 Introduction

The propagation of shock waves in a duct of non-uniform cross-sectional area or density was investigated by Chester (1954), Chisnell (1957), and Whitham (1958). The methods are generally referred to as the CCW theory. These authors used different approaches but obtained essentially similar end results. Chester considered small area variations of the duct and linearized the basic equations which are then be integrated to obtain the solution for the variation of the shock strength with area. Chisnell considered the wave interaction for a single step area change. He then integrated to obtain the result for a continuous area variation. Whitham simply replaced the variables along a $C+$ characteristic by the shocked state from the Rankine–Hugoniot equations to obtain a differential equation for the shock Mach number and area change, which can be integrated to give a shock Mach number–area relationship. The methods of the three authors are quite different and it is difficult to demonstrate their equivalence and relative merits. However, their final results are the same.

3.2 Chester's Theory

Chester considered the propagation of a shock wave in a variable area duct where for $x < 0$, the cross-sectional area is uniform, that is, $A = A_0 =$ constant and for $x > 0$, $A = A(x)$. Chester also assumed the rate of change of area $\frac{dA}{dx}$ to be small. The quasi-one-dimensional equations of motion for non-steady flow can be written as

$$\frac{\partial \rho}{\partial t} + \frac{\partial(\rho u)}{\partial x} + \frac{\rho u}{A}\frac{dA(x)}{dx} = 0, \tag{3.2.1}$$

$$\frac{\partial u}{\partial t} + u\frac{\partial u}{\partial x} + \frac{1}{\rho}\frac{\partial p}{\partial x} = 0, \tag{3.2.2}$$

$$\frac{Ds}{Dt} = 0, \tag{3.2.3}$$

where

$$\frac{D}{Dt} = \frac{\partial}{\partial t} + u\frac{\partial}{\partial x}.$$

For small area variation, that is, $A = A_0 + A'$ and $\frac{A'}{A_0} \ll 1$, the flow variables can be written as $p = p_1 + p'$, $\rho = \rho_1 + \rho'$, $u = u_1 + u'$, $s = s_1 + s'$ where the subscript 1 denotes the condition behind the initial shock wave in the uniform area section, $A = A_0 =$ constant. The equations for the flow perturbations can be obtained as

$$\frac{\partial \rho'}{\partial t} + \rho_1 \frac{\partial u'}{\partial x} + u_1 \frac{\partial \rho'}{\partial x} + \frac{\rho_1 u_1}{A_0} \frac{\mathrm{d}A'(x)}{\mathrm{d}x} = 0, \tag{3.2.4}$$

$$\frac{\partial u'}{\partial t} + u_1 \frac{\partial u'}{\partial x} + \frac{1}{\rho_1} \frac{\partial p'}{\partial x} = 0, \tag{3.2.5}$$

$$\left(\frac{\partial}{\partial t} + u_1 \frac{\partial}{\partial x} \right) s' = 0, \tag{3.2.6}$$

ρ' in Eq 3.2.4 can be replaced by p' using the isentropic relationship and we can write Eq 3.2.4 as

$$\frac{\partial p'}{\partial t} + u_1 \frac{\partial p'}{\partial x} + \rho_1 c_1^2 \frac{\partial u'}{\partial x} + \frac{\rho_1 c_1^2}{A_0} u_1 \frac{\mathrm{d}A'(x)}{\mathrm{d}x} = 0. \tag{3.2.7}$$

Expressing the entropy in terms of p and ρ, Eq 3.2.6 becomes

$$\left(\frac{\partial}{\partial t} + u_1 \frac{\partial}{\partial x} \right) \left(\rho' - \frac{1}{c_1^2} p' \right) = 0. \tag{3.2.8}$$

Writing Eqs 3.2.5 and 3.2.7 in characteristic form, we get

$$\left[\frac{\partial}{\partial t} + (u_1 \pm c_1) \frac{\partial}{\partial x} \right] \left(p' \pm \rho_1 c_1^2 u' \right) = -\frac{\rho_1 u_1 c_1^2}{A_0} \frac{\mathrm{d}A'(x)}{\mathrm{d}x}. \tag{3.2.9}$$

Integrating along the characteristics, we get along a C^+

$$p' + \rho_1 c_1 u' = -\frac{\rho_1 u_1 c_1^2}{u_1 + c_1} \frac{A'(x)}{A_0} + f_1 \left(x - (u_1 + c_1) t \right), \tag{3.2.10}$$

and along a C^-

$$p' - \rho_1 c_1 u' = -\frac{\rho_1 u_1 c_1^2}{u_1 - c_1} \frac{A'(x)}{A_0} + f_2 \left(x - (u_1 - c_1) t \right). \tag{3.2.11}$$

Integrating Eq 3.2.8 along the particle path $\frac{\mathrm{d}x}{\mathrm{d}t} = u_1$, we get

$$\left(\rho' - \frac{p'}{c_1^2} \right) = f_3(x - u_1 t). \tag{3.2.12}$$

The functions f_1, f_2, and f_3 are constants of integration which can be determined from initial and boundary conditions. Solving for p' and u' from Eqs 3.2.10 and 3.2.11 we obtain

$$p' = \frac{-\rho_1 u_1^2 c_1^2}{u_1^2 - c_1^2} \frac{A'(x)}{A_0} + \frac{1}{2} \left[f_1(x - (u_1 + c_1)t) + f_2(x - (u_1 - c_1)t) \right], \tag{3.2.13}$$

$$u' = \frac{u_1 c_1^2}{u_1^2 - c_1^2} \frac{A'(x)}{A_0} + \frac{1}{2\rho_1 c_1} \left[f_1(x - (u_1 + c_1)t) - f_2(x - (u_1 - c_1)t) \right]. \tag{3.2.14}$$

From the above equations, we note that the pressure and velocity perturbations depend on the local area change and on the perturbations carried along the C^- and C^- characteristics through the functions f_1 and f_2. Since the C^+ comes from the uniform region $x < 0$ where the perturbations are zero, $f_1 = 0$. The C^- carries the perturbations reflected from the shock front. Hence if the reflected C^- can influence the flow field significantly, then the influence will be carried back to the shock by C^+ and the f_1 will no longer be zero. However, the re-reflected perturbations are in general, higher order effects and can be neglected. Hence we may take $f_1 = 0$. Thus we may write Eq 3.2.10 as

$$p' + \rho_1 c_1 u' = -\frac{\rho_1 u_1 c_1^2}{u_1 + c_1} \frac{A'(x)}{A_0},$$

which give the pressure and velocity perturbations as a function of the local area change. Evaluating Eq 3.2.15 at the shock front, the Rankine–Hugoniot equations can be used to give

$$\frac{p_1}{p_0} = \frac{2\gamma}{\gamma + 1}\left(M_s^2 - \frac{\gamma - 1}{2\gamma}\right), \tag{3.2.15}$$

$$\frac{\rho_1}{\rho_0} = \frac{(\gamma + 1)M_s^2}{2 + (\gamma - 1)M_s^2}, \tag{3.2.16}$$

$$\frac{u_1}{u_0} = \frac{2}{\gamma + 1}\left(M_s - \frac{1}{M_s}\right), \tag{3.2.17}$$

$$\left(\frac{c_1}{c_0}\right)^2 = \frac{(2\gamma M_s^2 - (\gamma - 1))(2 + (\gamma - 1)M_s^2)}{(\gamma + 1)^2 M_s^2}. \tag{3.2.18}$$

The perturbations p', ρ', u' can be obtained by differentiating Eqs 3.2.15 to 3.2.17, that is,

$$p' = \frac{4\gamma p_0 M_s}{(\gamma + 1)} M_s', \tag{3.2.19}$$

$$u' = \frac{2c_0}{(\gamma + 1)}\left(1 + \frac{1}{M_s^2}\right) M_s', \tag{3.2.20}$$

$$\rho' = \frac{4(\gamma + 1)\rho_0 M_s M_s'}{(2 + (\gamma + 1)M_s^2)^2}, \tag{3.2.21}$$

where we have dropped the subscript 1 in the above equations.

Substituting Eqs 3.2.15 to 3.2.21 into Eq 3.2.15 and re-arranging yields

$$\frac{A'}{A_0} = \frac{-2M_s M_s'}{(M_s^2 - 1)K(M_s)}, \tag{3.2.22}$$

where

$$K(M_s) = 2\left\{\left(1 + \frac{1}{M_s^2} + 2\mu\right)\left(1 + \frac{2}{\gamma + 1}\frac{1 - \mu^2}{\mu}\right)\right\}^{-1}, \tag{3.2.23}$$

and

$$\mu^2 = \frac{2 + (\gamma - 1)M_s^2}{2\gamma M_s^2 - (\gamma - 1)}. \qquad (3.2.24)$$

Equation 3.2.22 gives the variation of the shock strength with area which can be integrated for a given initial shock Mach number and $A(x)$. Note that the determination of $M_s(x)$ does not require the solution for f_2 and f_3.

To complete the solution for the perturbations p', u', and ρ', we need to evaluate the functions f_2 and f_3. With $f_1 = 0$, Eq 3.2.13 becomes

$$p' + \frac{\rho_1 u_1^2 c_1^2 A'(x)}{(u_1^2 - c_1^2)A_0} = F_2(x - (u_1 - c_1)t),$$

where for convenience we have defined $F_2 = \frac{1}{2}f_2$. To evaluate at the shock front, we put $t = \frac{x}{D}$ where D is the shock velocity. The above equation becomes

$$F_2\left(x - (u_1 - c_1)\frac{x}{D}\right) = F_2\left(\frac{x}{k_1}\right) = p' + \frac{\rho_1 u_1^2 c_1^2}{(u_1^2 - c_1^2)}\frac{A'(x)}{A_0},$$

where

$$k_1 = \left(1 - \frac{u_1 - c_1}{D}\right)^{-1}.$$

Using the Rankine–Hugoniot equations for p' (Eq 3.2.19) and Eq 3.2.22 for $\dfrac{A'}{A_0}$, we obtain

$$F_2\left(\frac{x}{k_1}\right) = \frac{\rho_1 u_1^2 c_1^2}{(u_1^2 - c_1^2)}\frac{A'(x)}{A_0}\left(1 - \frac{(u_1^2 - c_1^2)}{u_1 c_1}K(M_s)\mu\right). \qquad (3.2.25)$$

From Eq 3.2.12 we obtain

$$f_3\left(x - \frac{u_1 x}{D}\right) = f_3\left(\frac{x}{k_3}\right) = \rho' - \frac{p'}{c_1^2},$$

where we have replaced $t = \frac{x}{D}$ and evaluated D at the shock. From Eq 3.2.19 for p', Eq 3.2.21 for ρ' and Eq 3.2.22 for M_s', we obtain

$$f_3\left(\frac{x}{k_3}\right) = \frac{\gamma - 1}{2}\frac{u_1^2}{c_1^2}\rho_1\left(1 - \frac{1}{M_s^2}\right)K(M_s)\frac{A_1}{A_0}, \qquad (3.2.26)$$

where $k_3 = \left(1 - \frac{u_1}{D}\right)^{-1}$.

With f_2 and f_3 determined, the solutions for the perturbation quantities p', u', and ρ' can be obtained as

$$p' = \frac{-\rho_1 u_1^2 c_1^2}{u_1^2 - c_1^2}\left\{K_2\frac{A(k_1(x - (u_1 - c_1)t)) - A_0}{A_0} - \frac{A(x) - A_0}{A_0}\right\}, \qquad (3.2.27)$$

where K_2 is defined as

$$K_2 = 1 - \frac{u_1^2 - c_1^2}{u_1 c_1} K(M_s)\mu,$$

$$u' = \frac{u_1 c_1^2}{u_1^2 - c_1^2} \left\{ \frac{u_1}{c_1} K_2 \frac{A(k_1(x - (u_1 - c_1)t)) - A_0}{A_0} - \frac{A(x) - A_0}{A_0} \right\}, \tag{3.2.28}$$

and

$$\rho' = \frac{1}{c_1^2} p' + K_4 \frac{A(k_3(x - u_1 t)) - A_0}{A_0}, \tag{3.2.29}$$

and K_4 is defined as

$$K_4 = \frac{\gamma - 1}{2} \frac{u_1^2}{c_1^2} \rho_1 \left(1 - \frac{1}{M_s^2}\right) K(M_s). \tag{3.2.30}$$

An important thing to note is that the perturbations generated at the shock propagate back at the local sound speed $u - c$ along the C^- characteristic. Thus when the flow becomes sonic, the perturbations remain stationary and thus build up and invalidate the linearized approximation. Friedman (1960) developed an improved version of Chester's solution by retaining the appropriate non-linear terms in the flow equations. However, it should be noted that when the perturbations build up, $u - c$ changes away from zero and disturbances will propagate upstream (or downstream) thus avoiding further accumulation. In general, one is interested in the shock Mach number–area variation and the detailed solutions for the perturbations behind the shock are of lesser importance.

3.3 Chisnell's Theory

Chisnell obtained the same result as Chester using a different approach. He first considered the propagation of a shock wave in a constant area duct in which an abrupt small area change is encountered. He then analyzed the wave interaction processes and solved for the perturbation of the shock velocity resulting from the area change. The various wave processes are illustrated in Fig. 3.1 and we have assumed that the flow behind the shock is supersonic so that the reflected wave is swept downstream of the area change. However, the analysis also applies if the flow is subsonic and the reflected wave propagates upstream of the area change. Across the contact surface separating regions 4 and 5, the pressure and particle velocity are continuous, that is, $p_4 = p_5$ and $u_4 = u_5$. The reflected wave is either a compression or an expansion wave depending on whether the area increases or decreases respectively. If the area change is small, that is, $\frac{A'}{A} \ll 1$, the reflected wave is weak and can be considered as an acoustic pulse, where, $p' = -\rho c u'$. Across the incident and transmitted shock waves, the Rankine–Hugoniot relationships link regions 1 and 2, and regions 6 and 5, respectively. Using the pressure ratio as the shock strength parameter, that is, $z = \frac{p_2}{p_1}$, instead of the shock Mach number,

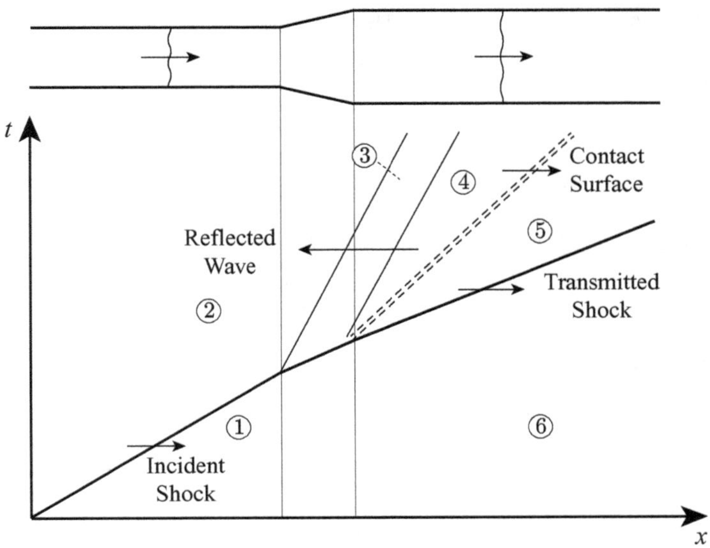

Figure 3.1 Various Wave Processes and the Flow Regions Between Them

the Rankine–Hugoniot equations linking the regions across the incident shock wave are given by

$$\rho_2 = \rho_1 \frac{(\gamma+1)z + (\gamma-1)}{(\gamma-1)z + (\gamma+1)}, \tag{3.3.1}$$

$$u_2 = (z-1)\left\{\frac{2p_1}{\rho_1[(\gamma-1) + z(\gamma+1)]}\right\}^{\frac{1}{2}}, \tag{3.3.2}$$

$$M_2 = (z-1)\left\{\frac{2}{\gamma z[(\gamma+1) + z(\gamma-1)]}\right\}^{\frac{1}{2}}, \tag{3.3.3}$$

where $M_2 = \frac{u_2}{c_2}$ and $c_2^2 = \frac{\gamma p_2}{\rho_2}$. Across the area change, Chisnell assumed steady quasi-one-dimensional isentropic flow. Thus between regions 2 and 3, we write

$$\frac{p_3}{p_2} = 1 - \frac{\gamma M_2^2}{M_2^2 - 1}\frac{A'}{A}, \tag{3.3.4}$$

$$\frac{\rho_3}{\rho_2} = 1 - \frac{M_2^2}{M_2^2 - 1}\frac{A'}{A}, \tag{3.3.5}$$

$$\frac{u_3}{u_2} = 1 + \frac{1}{M_2^2 - 1}\frac{A'}{A}. \tag{3.3.6}$$

Since $p_1 = p_6$ and across the interface $p_4 = p_5$, we can write the pressure ratio across the transmitted shock $\frac{p_5}{p_6} = \frac{p_4}{p_1}$. Since the strength of the transmitted shock differs slightly from the incident shock, we may write $\frac{p_5}{p_6}$ as $z + z' = z\left(1 + \frac{z'}{z}\right)$ where

$\frac{z'}{z} \ll 1$. Thus $\frac{p_5}{p_6} = \frac{p_4}{p_1} = z\left(1 + \frac{z'}{z}\right)$. The pressure $\frac{p_3}{p_1}$ can be written as $\frac{p_3}{p_1} = \frac{p_3}{p_2}\frac{p_2}{p_1} = z\left(1 - \frac{\gamma M_2^2}{M_2^2-1}\frac{A'}{A}\right)$ where we have used Eq 3.3.4 for $\frac{p_3}{p_2}$. We now write

$$\frac{p_4}{p_3} = \frac{\frac{p_4}{p_1}}{\frac{p_3}{p_1}} = \frac{z\left(1 + \frac{z'}{z}\right)}{z\left(1 - \frac{\gamma M_2^2}{M_2^2-1}\frac{A'}{A}\right)}.$$

Expanding and retaining only first-order terms yield

$$\frac{p_4}{p_3} = 1 + \frac{z'}{z} + \frac{\gamma M_2^2}{M_2^2 - 1}\frac{A'}{A} + \cdots \tag{3.3.7}$$

A similar expression for $\frac{u_4}{u_3}$ can be obtained as follows: Across the transmitted shock, the Rankine–Hugoniot equations give

$$u_5 = u_4 = (z + z' - 1)\left\{\frac{2p_1}{\rho_1[(\gamma - 1) + (z + z')(\gamma + 1)]}\right\},$$

where we have used $z + z'$ for the strength of the transmitted shock and we also write $p_6 = p_1$ and $\rho_6 = \rho_1$. Across the incident shock, u_2 is given by Eq 3.3.2 and across the area change u_3 is given by Eq 3.3.6. Thus we write

$$\frac{u_4}{u_3} = \frac{(z-1)\left(1 + \frac{z'}{z-1}\right)\left\{\frac{2p_1}{\rho_1[(\gamma-1)+(z+z')(\gamma+1)]}\right\}^{\frac{1}{2}}}{(z-1)\left[\frac{2p_1}{[\rho_1(\gamma-1)+z(\gamma+1)]}\right]\left(1 + \frac{1}{M_2^2-1}\frac{A'}{A}\right)},$$

$$= \left(1 + \frac{z'}{z-1}\right)\left(1 - \frac{1}{M_2^2 - 1}\frac{A'}{A} + \cdots\right)\left(1 - \frac{1}{2}\frac{(\gamma+1)z'}{[(\gamma - 1) + z(\gamma + 1)]} + \cdots\right),$$

and retaining only first-order terms yield

$$\frac{u_4}{u_3} = 1 + \frac{z'}{z-1} - \frac{1}{M_2^2 - 1}\frac{A'}{A} - \frac{(\gamma+1)z'}{2[(\gamma - 1) + z(\gamma + 1)]}. \tag{3.3.8}$$

Assuming a simple wave separating regions 3 and 4, we can write the Riemann invariant linking the two regions as

$$\frac{2}{\gamma - 1}c_3 + u_3 = \frac{2}{\gamma - 1}c_4 + u_4,$$

and obtain

$$c_3 - c_4 = c_3\left(1 - \frac{c_4}{c_3}\right) = \frac{\gamma - 1}{2}(u_4 - u_3).$$

Using the isentropic relationship, that is,

$$\frac{c_4}{c_3} = \left(\frac{p_4}{p_3}\right)^{\frac{\gamma-1}{2\gamma}} = 1 - \frac{\gamma - 1}{2}\left(\frac{u_4 - u_3}{c_3}\right),$$

we obtain

$$\frac{p_4}{p_3} = \left\{1 - \frac{\gamma - 1}{2}\left(\frac{u_4 - u_3}{c_3}\right)\right\}^{\frac{2\gamma}{\gamma-1}} \simeq 1 - \frac{\gamma u_3}{c_3}\left(\frac{u_4}{u_3} - 1\right) + \cdots$$

where we have expanded the bracketed form for small velocity differences. Thus we obtain

$$\frac{p_4}{p_3} \simeq 1 - \gamma M_3 \left(\frac{u_4}{u_3} - 1\right).$$

Substituting the expressions for $\frac{p_4}{p_3}$ (Eq 3.3.7) and $\frac{u_4}{u_3}$ (Eq 3.3.8) into the above, we obtain

$$\frac{z'}{z} + \frac{\gamma M_2^2}{M_2^2 - 1}\frac{A'}{A} = -\gamma M_3 \left(\frac{z'}{z} - \frac{(\gamma+1)z'}{2[(\gamma-1) + z(\gamma+1)]} - \frac{1}{M_2^2 - 1}\frac{A'}{A}\right).$$

Since M_3 and M_2 differ by a term of the order of $\frac{A'}{A}$, we may replace M_3 by M_2 which is given by the Rankine–Hugoniot equation (Eq 3.3.2) as

$$M_3 \simeq M_2 = (z-1)\left[\frac{2}{\gamma z[(\gamma+1) + z(\gamma-1)]}\right]^{\frac{1}{2}}.$$

Taking the limit $\frac{A'}{A}$ as $A' \to 0$, we get a differential equation for $\frac{\mathrm{d}A}{\mathrm{d}z}$ as

$$\frac{1}{A}\frac{\mathrm{d}A}{\mathrm{d}z} = \frac{1}{\gamma z} + \frac{1}{z-1} - \frac{\gamma+1}{2[(\gamma+1)z + (\gamma-1)]} + \left[\frac{2}{\gamma z[(\gamma-1)z + (\gamma+1)]}\right]^{\frac{1}{2}}$$
$$\left[1 - \frac{(\gamma+1)(z-1)}{2[(\gamma+1)z + (\gamma-1)]} + \frac{(\gamma-1)z + (\gamma+1)}{2(z-1)}\right]. \tag{3.3.9}$$

If we replace z by the Rankine–Hugoniot equation

$$z = \frac{2\gamma M_s^2 - (\gamma-1)}{\gamma+1}. \tag{3.3.10}$$

Equation 3.3.9 can be reduced to the form given previously by Chester, that is,

$$\frac{A'}{A} = \frac{-2M_s M_s'}{(M_s^2 - 1)K(M_s)}.$$

Chisnell managed to integrate Eq 3.3.9 and obtained the shock-strength-area relationships as

$$Af(z) = \text{constant}, \tag{3.3.11}$$

where the function $f(z)$ is defined as

$$f(z) = z^{\frac{1}{\gamma}}(z-1)\left(z + \frac{\gamma-1}{\gamma+1}\right)^{-\frac{1}{2}}\left(\frac{1+R}{1-R}\right)^{\sqrt{\frac{\gamma}{2(\gamma-1)}}}$$
$$\left[\frac{R - \left(\frac{\gamma-1}{2\gamma}\right)^{\frac{1}{2}}}{R + \left(\frac{\gamma-1}{2\gamma}\right)^{\frac{1}{2}}}\right] \exp\left[\left(\frac{2}{\gamma-1}\right)^{\frac{1}{2}} \tan^{-1}\left(\frac{2\gamma^{\frac{1}{2}}R}{\gamma-1}\right)\right], \tag{3.3.12}$$

where

$$R = \left[1 + \frac{\gamma+1}{(\gamma-1)z}\right],$$

and

$$z(M_s) = \frac{2\gamma}{\gamma+1} M_s^2 - \frac{\gamma-1}{\gamma+1}.$$

The constant of integration in Eq 3.3.11 can be evaluated given the initial shock strength $z = z_0$ at $A = A_0$.

It should be noted that Chisnell assumed steady flow across the area change. In reality, steady state requires a finite relaxation time after the passage of the shock wave. In applying Chisnell's analysis to a continuous area change, the reflected waves will interact with the area change and the re-reflected waves will overtake the shock and influence its motion. Thus, we have to assume that the re-reflected waves cause only second-order effects for small area change. Thus, similar to Chester's theory, the shock strength becomes a function of the local area change only.

3.4 Whitham's Theory

Whitham (1958) obtained the same area–Mach number relationship of Chester and Chisnell in a more straightforward manner. He wrote the non-linear differential equation for the C^+ characteristic and substituted the expressions for the variables p, ρ, u, and c in terms of M_s from the Rankine–Hugoniot equations. Integrating the resultant equation then gives the variation of the shock Mach number with the area change $A(x)$.

The equation for a C^+ characteristic is given by

$$\mathrm{d}p + \rho u \mathrm{d}u + \frac{\rho^2 c u}{u+c}\frac{\mathrm{d}A}{A} = 0, \tag{3.4.1}$$

where $\frac{\mathrm{d}x}{\mathrm{d}t} = u + c$ for a C^+. The Rankine–Hugoniot relationships for a normal shock are given by

$$\frac{p_1}{p_0} = \frac{2\gamma M_s^2 - (\gamma-1)}{\gamma+1}, \tag{3.4.2}$$

$$\frac{\rho_1}{\rho_0} = \frac{(\gamma+1)M_s^2}{2+(\gamma-1)M_s^2}, \tag{3.4.3}$$

$$\frac{u_1}{c_0} = \frac{2}{\gamma+1}\left(M_s - \frac{1}{M_s}\right). \tag{3.4.4}$$

If we drop the subscript 1 and take the differential for $\mathrm{d}p$ and $\mathrm{d}u$, we get

$$\frac{\mathrm{d}p}{p_0} = \frac{4\gamma M_s\,\mathrm{d}M_s}{\gamma+1}, \tag{3.4.5}$$

$$\frac{\mathrm{d}u}{c_0} = \frac{2}{\gamma+1}\left(\frac{M_s^2+1}{M_s^2}\right). \tag{3.4.6}$$

The sound speed across the shock is given by

$$\left(\frac{c_1}{c_0}\right)^2 = \frac{[2\gamma M_s^2 - (\gamma-1)][2+(\gamma-1)M_s^2]}{(\gamma+1)^2 M_s^2}. \tag{3.4.7}$$

Substituting Eqs 3.4.2–3.4.7 into Eq 3.4.1 and re-arranging yields

$$\frac{dA}{A} = -\frac{\lambda(M_s)\,dM_s}{M_s^2}, \tag{3.4.8}$$

where

$$\lambda(M_s) = \left(1 + \frac{2}{\gamma+1}\frac{1-\mu^2}{\mu}\right)\left(1 + 2\mu + \frac{1}{M_s}\right),$$

and

$$\mu^2 = \frac{2 + (\gamma-1)M_s^2}{2\gamma M_s^2 - (\gamma-1)}.$$

The function $\lambda(M_s)$ is essentially Chester's function $K(M_s)$, that is,

$$\lambda(M_s) = \frac{2}{K(M_s)}.$$

Whitham's choice for $\lambda(M_s)$ is perhaps based on its weak dependence on M_s. For $M_s \to 1$, Whitham showed that $\lambda(M_s) \to 4$, whereas for $M_s \to \infty$, $\lambda \to 1 + \frac{2}{\gamma} + \sqrt{\frac{2\gamma}{\gamma-1}}$ which for $\gamma = 1.4$, $\lambda \to 5.07$. Thus we see that for the range $1 < M_s < \infty$, the range for $\lambda(M_s)$ is only $4 < \lambda < 5.07$. Thus we may take $\lambda(M_s)$ to be constant and Eq 3.4.8 then integrates to yield

$$A(M_s^2 - 1)^{\frac{\lambda}{2}} = \text{constant}.$$

For a weak shock where $M_s \simeq 1$, $\lambda \simeq 4$, then $M_s^2 - 1 \sim A^{-\frac{1}{2}}$. For the spherical geometry where $A \sim x^2$, $M_s^2 - 1 \sim \frac{1}{x}$ and since the shock overpressure $\frac{\Delta p}{p_0} \sim M_s^2 - 1$, the result gives $\frac{\Delta p}{p_0} \sim \frac{1}{x}$ in accord with acoustic theory. For strong shocks where $M_s \to \infty$ and $\lambda \to 5.07$, we obtain,

$$A(M_s^2 - 1)^{\frac{5.07}{2}} = \text{constant}.$$

Thus $A(M_s^2 - 1)^{2.5} = \text{constant}$. For cylindrical implosions where $A \sim x_s$ and $M_s \gg 1$, then $x_s M_s^{5.07} = \text{constant}$ or $M_s \sim x_s^{-\frac{1}{5.07}} \sim x_s^{-0.19723}$. For spherical implosions, $A \sim x_s^2$ and $x_s^2 M_s^{5.07} = \text{constant}$ which gives $M_s \sim x_s^{-0.39447}$. The exact similarity solution of Guderley (1942) for strong converging shock waves gives the exponent for x_s as 0.19723 for the cylindrical case and 0.39447 for the spherical case. Thus the agreement between the simple area rule of Whitham and the exact similarity solution is amazingly good. The area–Mach number rule does not apply for diverging blast waves because the shock strength depends on the flow field behind it and not just on the local area change.

Whitham's theory can also be applied to describe the propagation of a shock wave in a non-uniform density medium, that is, $\rho_0(x)$. If $\rho_0(x)$ is decreasing in the direction of the shock motion, the shock amplifies. If $\rho_0(x)$ is increasing in the direction of the shock motion, then the shock attenuates. We shall consider a strong shock (i.e., $M_s \to \infty$) and the shocked state takes on its limiting form, that is,

$$\left.\begin{aligned} \frac{p_1}{\rho_0 D^2} &= \frac{2}{\gamma+1} = \frac{u_1}{D} \\ \frac{\rho_1}{\rho_0} &= \frac{\gamma+1}{\gamma-1} \end{aligned}\right\} \tag{3.4.9}$$

Table 3.1 Comparisons Between Whitham's Theory and Similarity Solution

w	Whitham	Similarity
0.5	−0.107625	−0.103516
1.0	−0.215250	−0.202135
1.5	−0.323876	−0.290388

For the planar geometry the equation for the C^+ becomes

$$\mathrm{d}p + \rho c \mathrm{d}u = 0,$$

and using Whitham's "recipe" and applying it to the shock, we write

$$\mathrm{d}p_1 + \rho_1 c_1 \mathrm{d}u_1 = 0.$$

Using the strong shock conditions given by Eqs 3.4.9, we obtain

$$\mathrm{d}p_1 = \frac{2}{\gamma+1}(D^2 \mathrm{d}\rho_0 + 2D\rho_0 \mathrm{d}D),$$

$$\mathrm{d}u_1 = \frac{2}{\gamma+1}\mathrm{d}D,$$

and

$$c_1^2 = \frac{\gamma p_1}{\rho_1} = \frac{2\gamma(\gamma-1)D^2}{(\gamma+1)^2}.$$

Substituting the above into the characteristic equation yields

$$\frac{\mathrm{d}D}{D} = -\frac{1}{2+\sqrt{\frac{2\gamma}{\gamma-1}}}\frac{\mathrm{d}\rho_0}{\rho_0}. \tag{3.4.10}$$

Equation 3.4.10 indicates that $D \sim \rho_0^{-\alpha}$, where

$$\alpha = \left(2+\sqrt{\frac{2\gamma}{\gamma-1}}\right)^{-1},$$

and the shock pressure $p_1 \sim \rho_0^{1-2x}$. For $\gamma = 1.4$, the value of $\alpha = 0.21525$. If the density variation is given by a power law $\rho_0(x) \sim x^w$, then $D \sim x^{-w\alpha}$. For strong shocks propagating in a non-uniform density field, there exists also an exact similarity solution. Comparing the result from Whitham's theory to the similarity solution for $\gamma = 1.4$, the various values of w are given in Table 3.1.

As can be observed, the results are in reasonable agreement in spite of the simplicity of Whitham's method.

4 Blast Wave Theory

4.1 Introduction

A blast wave is a shock wave whose strength decreases as it propagates away from the source. Blast waves are generally formed by the rapid release of energy from a concentrated source (e.g., detonation of a condensed explosive charge). At early times, the blast motion is influenced by the initial properties of the source, for example, source dimension and energy release profile. At later times, the shock radius and time are large as compared to the source dimension and the characteristic time of the energy release. The source characteristic now no longer influences the shock motion. Thus the blast wave can be approximated as that due to an instantaneous point energy source. The propagation of an ideal strong point blast is described by an exact self-similar solution. This solution was first obtained independently in the early 1940s by Taylor (1950b,c), von Neumann (1963), and Sedov (1959). Noting that the important parameter of the problem (i.e., E_0, ρ_0, R_s, t) can form only one dimensionless group $\left(\frac{E_0}{\rho_0}\frac{t^2}{R_s^5}\right)^{1/5}$ Taylor (1950b,c) obtained the result for the shock position as

$$R_s = A\left(\frac{E_0}{\rho_0}\right)^{\frac{1}{5}} t^{\frac{2}{5}}, \tag{4.1.1}$$

where A is a constant which can be determined from the conservation of energy. Equation 4.1.1 can also be written in the form

$$\frac{5}{2}\log_{10} R_s = B + \frac{1}{2}\log_{10}\left(\frac{E_0}{\rho_0}\right) + \log_{10} t. \tag{4.1.2}$$

The above equation gives a straight line in the $\frac{5}{2}\log_{10} R_s$ versus $\log_{10} t$ plot and Fig. 4.1 illustrates the remarkable agreement with the experimental data from the high speed movie of the fireball from the Alamogordo test of the first nuclear bomb.

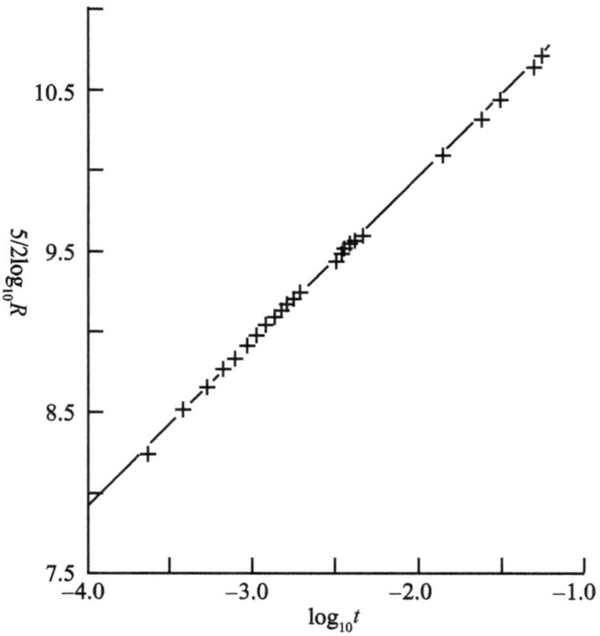

Figure 4.1 Logarithmic Plot of the Blast Radius Versus Time. *Source:* Taylor (1950a)

4.2 Basic Equations

The basic equations that govern the flow behind the blast wave are given as

$$\frac{\partial \rho}{\partial t} + u\frac{\partial \rho}{\partial r} + \rho\frac{\partial u}{\partial r} + \frac{j\rho u}{r} = 0, \tag{4.2.1}$$

$$\frac{\partial u}{\partial t} + u\frac{\partial u}{\partial r} + \frac{1}{\rho}\frac{\partial p}{\partial r} = 0, \tag{4.2.2}$$

$$\frac{\mathrm{D}s}{\mathrm{D}t} = 0, \tag{4.2.3}$$

where $j = 0, 1, 2$ for planar, cylindrical, and spherical symmetry, respectively.

From thermodynamics, the entropy can be expressed as

$$T\frac{\mathrm{D}s}{\mathrm{D}t} = \frac{\mathrm{D}e}{\mathrm{D}t} - \frac{p}{\rho^2}\frac{\mathrm{D}\rho}{\mathrm{D}t}.$$

Noting that $e = \frac{p}{(\gamma-1)\rho}$ for a perfect gas and using Eq 4.2.1 for $\frac{\mathrm{D}\rho}{\mathrm{D}t}$, Eq 4.2.3 becomes

$$\frac{\mathrm{D}p}{\mathrm{D}t} + \gamma p\frac{\partial u}{\partial r} + \frac{\gamma j p u}{r} = 0. \tag{4.2.4}$$

We seek a self-similar solution to Eqs 4.2.1–4.2.3 of the form where the dependent variables are functions of r, t via the similarity variable $\xi = \frac{r}{R_s(t)}$. Defining the following non-dimensional variables,

$$\psi(\xi) = \frac{\rho(r,t)}{\rho_0}, \quad \phi(\xi) = \frac{u(r,t)}{\dot{R}_s(t)}, \quad f(\xi) = \frac{p(r,t)}{\rho_0\dot{R}_s^2(t)}, \quad \xi = \frac{r}{R_s(t)}, \tag{4.2.5}$$

Equations 4.2.1, 4.2.2, and 4.2.4 become

$$(\phi-\xi)\psi'+\psi\phi'+\frac{j\phi\psi}{\xi}=0, \tag{4.2.6}$$

$$(\phi-\xi)\phi'+\theta\phi+\frac{1}{\psi}f'=0, \tag{4.2.7}$$

$$(\phi-\xi)f'+2\theta f+\gamma f\phi'+\frac{\gamma j\phi f}{\xi}=0, \tag{4.2.8}$$

where the parameter θ is given by

$$\theta=\frac{R_s\ddot{R}_s}{\dot{R}_s^2}. \tag{4.2.9}$$

For a self-similar solution, θ in Eqs 4.2.7 and 4.2.8 must not be dependent on R_s (or time) explicitly. Thus

$$\theta=\frac{\ddot{R}_sR_s}{\dot{R}_s^2}=\text{constant}.$$

Rewriting Eq 4.2.9 as

$$\theta=\frac{R_s\ddot{R}_s}{\dot{R}_s^2}=\frac{\mathrm{d}\ln\dot{R}_s}{\mathrm{d}\ln R_s}=\text{constant},$$

we can integrate and obtain

$$\dot{R}_s=CR_s^{\theta}, \tag{4.2.10}$$

where C is a constant of integration.

Equations 4.2.6–4.2.8 can be integrated when θ is determined. Solving for the derivatives ψ', ϕ', f', we obtain

$$\psi'=\frac{-\left[(\phi-\xi)^2\frac{j\phi\psi}{\xi}+2\theta f-\theta\phi\psi(\phi-\xi)\right]}{(\phi-\xi)\left[(\phi-\xi)^2-\frac{\gamma f}{\psi}\right]}, \tag{4.2.11}$$

$$\phi'=\frac{-\theta\phi(\phi-\xi)+\frac{f}{\psi}\left(2\theta+\frac{\gamma j\phi}{\xi}\right)}{(\phi-\xi)^2-\frac{\gamma f}{\psi}}, \tag{4.2.12}$$

$$f'=\frac{-(\phi-\xi)\left(2\theta f+\frac{\gamma j\phi f}{\xi}\right)+\gamma f\phi\theta}{(\phi-\xi)^2-\frac{\gamma f}{\psi}}. \tag{4.2.13}$$

The boundary conditions at the shock front $\xi=1$ are given by the Rankine–Hugoniot equations. For strong shocks, we write

$$\psi(1)=\frac{\rho_1}{\rho_0}=\frac{\gamma+1}{\gamma-1}, \tag{4.2.14}$$

$$\phi(1)=\frac{u_1}{\dot{R}_s}=\frac{2}{\gamma+1}, \tag{4.2.15}$$

$$f(1)=\frac{p_1}{\rho_0\dot{R}_s^2}=\frac{2}{\gamma+1}. \tag{4.2.16}$$

The solution must also satisfy the rear boundary condition that at the center of symmetry $r = 0$ ($\xi = 0$), the particle velocity must vanish, that is, $\phi(0) = 0$.

4.3 The Energy Integral

To determine θ, we consider the energy integral. At any instant when the shock front is at R_s, the conservation of energy can be written as

$$E_0 = \int_0^{R_s} \left(\rho e + \frac{\rho u^2}{2} \right) k_j r^j \mathrm{d}r, \tag{4.3.1}$$

where $k_j = 1, 2\pi, 4\pi$ for the three geometries of planar ($j = 0$), cylindrical ($j = 1$), and spherical ($j = 2$). In Eq 4.3.1, we have neglected the initial internal energy e_0 when compared to e. For a perfect gas, $e = \frac{p}{\rho(\gamma-1)}$ and, in terms of the non-dimensional variables ψ, ϕ, f, ξ, Eq 4.3.1 can be written as

$$E_0 = \rho_0 k_j I \dot{R}_s^2 R_s^{j+1}, \tag{4.3.2}$$

where I is the dimensionless integral

$$I = \int_0^1 \left(\frac{f}{\gamma - 1} + \frac{\psi\phi^2}{2} \right) \xi^j \mathrm{d}\xi. \tag{4.3.3}$$

From Eq 4.2.10 where $\dot{R}_s = CR_s^{\theta}$, we can integrate and obtain

$$R_s = [C(1-\theta)]^{\frac{1}{1-\theta}} t^{\frac{1}{1-\theta}}. \tag{4.3.4}$$

We can write the above in a more convenient form as

$$R_s = At^N, \tag{4.3.5}$$

where

$$A = [C(1-\theta)]^{\frac{1}{1-\theta}}, \tag{4.3.6}$$

$$N = \frac{1}{1-\theta}, \tag{4.3.7}$$

or

$$\theta = -\left(\frac{1-N}{N} \right). \tag{4.3.8}$$

Using Eq 4.2.10, the energy integral becomes

$$E_0 = \rho_0 k_j I \dot{R}_s^2 R_s^{j+1} = \rho_0 k_j I C^2 R_s^{2\theta+j+1}. \tag{4.3.9}$$

Since the left hand is a constant, the exponent of R_s on the right hand side must vanish, and hence we obtain

$$2\theta + j + 1 = 0,$$

thus

$$\theta = -\left(\frac{j+1}{2} \right). \tag{4.3.10}$$

Equation 4.3.7 then gives

$$N = \frac{1}{1-\theta} = \frac{2}{j+3}. \tag{4.3.11}$$

Thus $R_s \sim t^{2/3}$, $R_s \sim t^{1/2}$, and $R_s \sim t^{2/5}$, for strong planar, cylindrical, and spherical blast waves, respectively. From Eq 4.3.9, the constant C can be obtained as

$$C = \left(\frac{E_0}{\rho_0}\frac{1}{k_j I}\right)^{\frac{1}{2}}, \tag{4.3.12}$$

and substituting the above equation into Eq 4.2.10, we obtain

$$\dot{R}_s^2 = \frac{E_0}{\rho_0}\frac{1}{k_j I} R_s^{-(j+1)}. \tag{4.3.13}$$

Defining a characteristic explosion length R_0 as

$$R_0 = \left(\frac{E_0}{p_0}\right)^{\frac{1}{j+1}}, \tag{4.3.14}$$

and noting that $M_s = \frac{\dot{R}_s}{c_0}$, Eq 4.3.13 can be written as

$$M_s^2 = \left(\frac{1}{\gamma k_j I}\right)\left(\frac{R_s}{R_0}\right)^{-(j+1)}. \tag{4.3.15}$$

From Eq 4.2.16, we write the shock pressure ratio as

$$\frac{p_1}{p_0} = \gamma M_s^2 \frac{2}{\gamma+1},$$

and using Eq 4.3.15, the shock pressure becomes

$$\frac{p_1}{p_0} = \frac{2}{\gamma+1}\left(\frac{1}{k_j I}\right)\left(\frac{R_s}{R_0}\right)^{-(j+1)}. \tag{4.3.16}$$

From Eqs 4.3.15 and 4.3.16, we note that M_s^2 and the shock pressure are proportional to the averaged volumetric energy density of the blast wave. From Eqs 4.3.6 and 4.3.12, the constant A can be obtained as

$$A = \left(\frac{E_0}{\rho_0}\right)^{\frac{1}{j+3}}\left(\frac{(j+3)^2}{4k_j I}\right)^{\frac{1}{j+3}}, \tag{4.3.17}$$

and the shock trajectory given by Eq 4.3.5 then becomes

$$\frac{R_s}{R_0} = \left[\frac{(j+3)^2}{4\gamma k_j I}\right]^{\frac{1}{j+3}}\left(\frac{c_0 t}{R_0}\right)^{\frac{2}{j+3}}. \tag{4.3.18}$$

The shock temperature ratio $\frac{T_1}{T_0}$ can also be obtained by noting $c_1^2 = \frac{\gamma p_1}{\rho_1}$ and hence $\frac{T_1}{T_0} = \left(\frac{c_1}{c_0}\right)^2 = \frac{p_1/p_0}{\rho_1/\rho_0}$. Using Eq 4.3.16 for the shock pressure and $\frac{\rho_1}{\rho_0} = \frac{\gamma+1}{\gamma-1}$, the shock temperature is obtained as

$$\frac{T_1}{T_0} = \left[\frac{2(\gamma-1)}{(\gamma+1)^2}\frac{1}{k_j I}\right]\left(\frac{R_s}{R_0}\right)^{-(j+1)}. \tag{4.3.19}$$

Similarly, the particle velocity behind the shock can also be determined as

$$\frac{u_1}{c_0} = \left(\frac{2}{\gamma+1}\right)\left(\frac{1}{\gamma k_j I}\right)^{\frac{1}{2}}\left(\frac{R_s}{R_0}\right)^{-\frac{j+1}{2}}. \tag{4.3.20}$$

All these expressions (i.e., Eqs 4.3.15, 4.3.16, 4.3.18–4.3.20) contain the dimensionless integral I (Eq 4.3.3) which can be evaluated when the solutions for $\psi(\xi)$, $\phi(\xi)$ and, $f(\xi)$ are determined upon the integration of the basic equations (i.e., Eqs 4.2.11–4.2.13). von Neumann, Kynch, and others have obtained closed form analytical solutions for the constant energy strong blast problem for the particular value of $\theta = -\left(\frac{j+1}{2}\right)$. However, the closed form solution is rather lengthy and complicated and it is more convenient to simply integrate Eqs 4.2.11–4.2.13 numerically.

Note that an alternate definition of the explosion length was given by Bach and Lee (1969) which includes the geometrical factors k_j and γ, that is,

$$R_0 = \left(\frac{E_0}{\gamma p_0 k_j}\right)^{\frac{1}{j+1}} = \left(\frac{E_0}{\rho_0 c_o^2 k_j}\right)^{\frac{1}{j+1}}. \tag{4.3.21}$$

This definition has the advantage that k_j no longer appears in the various expressions and thus simplifying Eqs 4.3.15–4.3.20. In terms of the R_0 defined by Eqs 4.3.21, 4.3.15, 4.3.16, and 4.3.18 reduce to the following

$$M_s^2 = \frac{1}{I}\left(\frac{R_s}{R_0}\right)^{-(j+1)}, \tag{4.3.22}$$

$$\frac{p_1}{p_0} = \frac{2\gamma}{\gamma+1}\frac{1}{I}\left(\frac{R_s}{R_0}\right)^{-(j+1)}, \tag{4.3.23}$$

$$\frac{R_s}{R_0} = \left(\frac{(j+3)^2}{4I}\right)^{\frac{1}{j+3}}\left(\frac{c_0 t}{R_0}\right)^{\frac{2}{j+3}}. \tag{4.3.24}$$

The new definition of R_0 given by Eq 4.3.21 also represents the ratio of the blast energy E_0 to the initial internal energy inside a volume of dimension R_0. This ratio is a constant of the order of unity, that is,

$$\frac{E_0}{\frac{\rho_0 k_j R_0^{j+1} e_0}{j+1}} = \gamma(\gamma-1)(\gamma+1) \simeq o[1], \tag{4.3.25}$$

where we have used the expression $e_0 = \frac{p_0}{\rho_0(\gamma-1)} = \frac{c_0^2}{\gamma(\gamma-1)}$ for a perfect gas.

For $\gamma = 1.4$ and the value of $\theta = -\left(\frac{j+1}{2}\right)$, Eqs 4.2.11–4.2.13 are integrated numerically for planar ($j = 0$), cylindrical ($j = 1$), and spherical ($j = 2$) blast waves. The strong shock boundary conditions given by Eqs 4.2.14–4.2.16 are used at the shock front ($\xi = 1$). The velocity, density, and pressure profiles are shown in Figs. 4.2–4.4. The temperature profile behind the blast wave can be obtained from the pressure and density profiles (i.e., $\frac{T}{T_1} = \frac{P}{P_1}\frac{\rho_1}{\rho} = \frac{f}{f_1}\frac{\psi_1}{\psi}$), and is shown in Fig. 4.5.

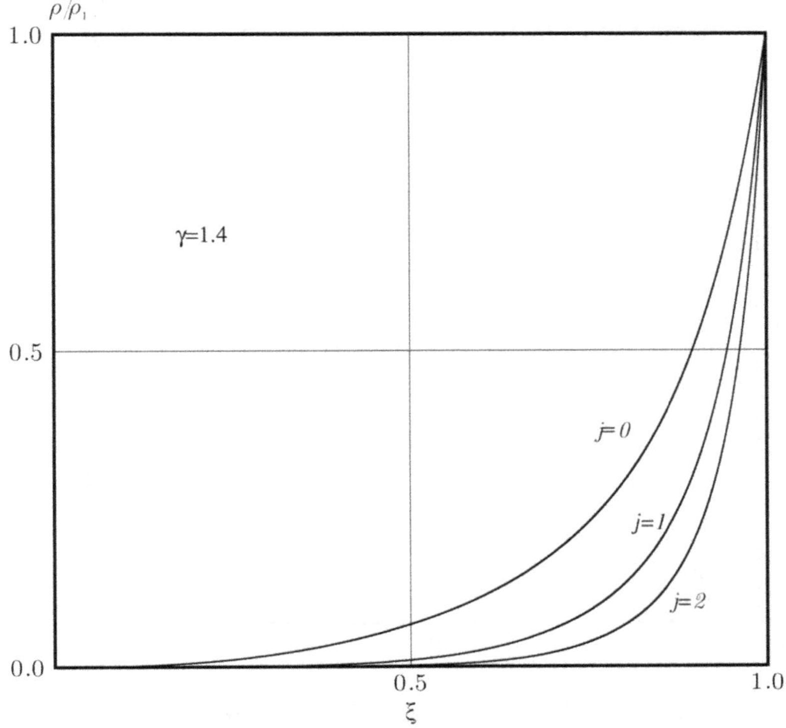

Figure 4.2 Density Distribution Behind the Shock Wave

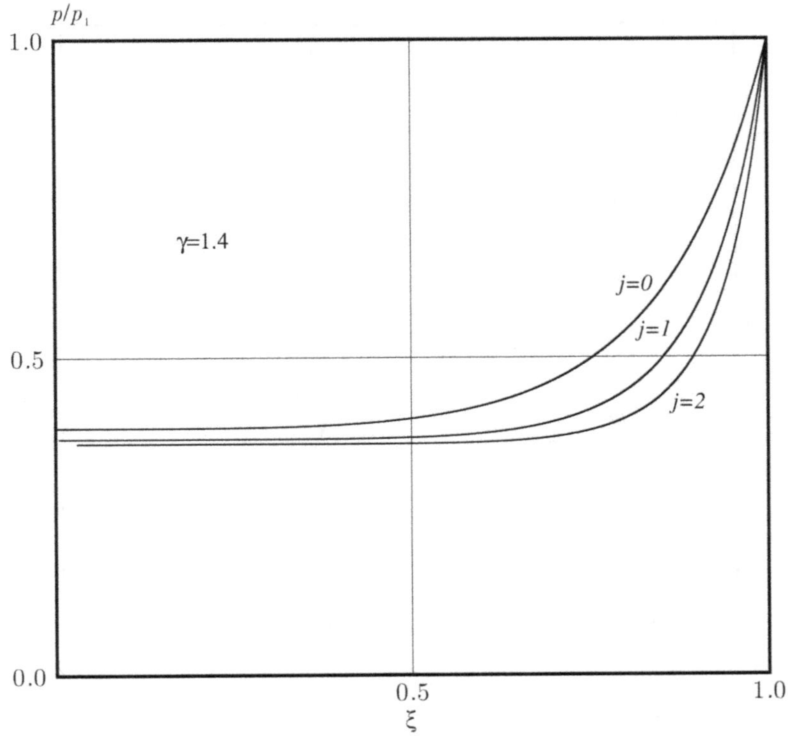

Figure 4.3 Pressure Distribution Behind the Shock Wave

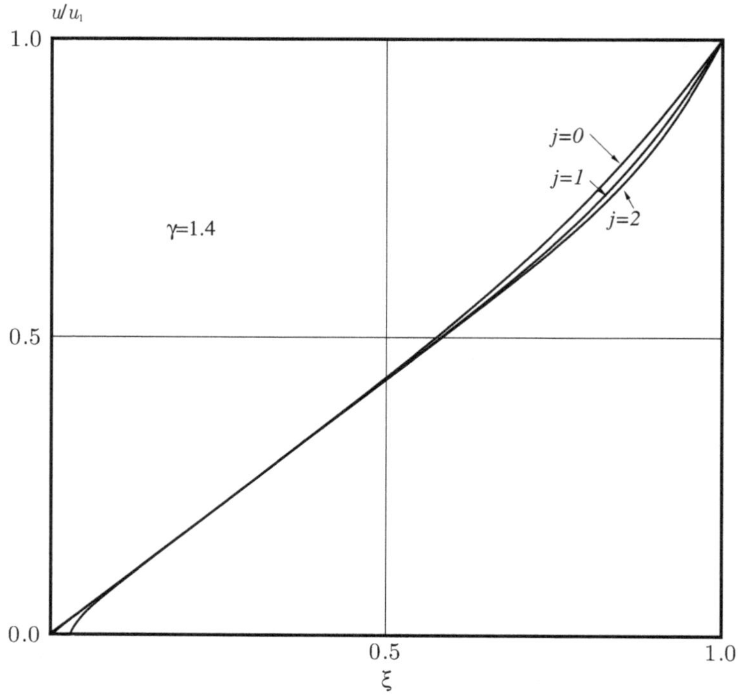

Figure 4.4 Particle Velocity Distribution Behind the Shock Wave

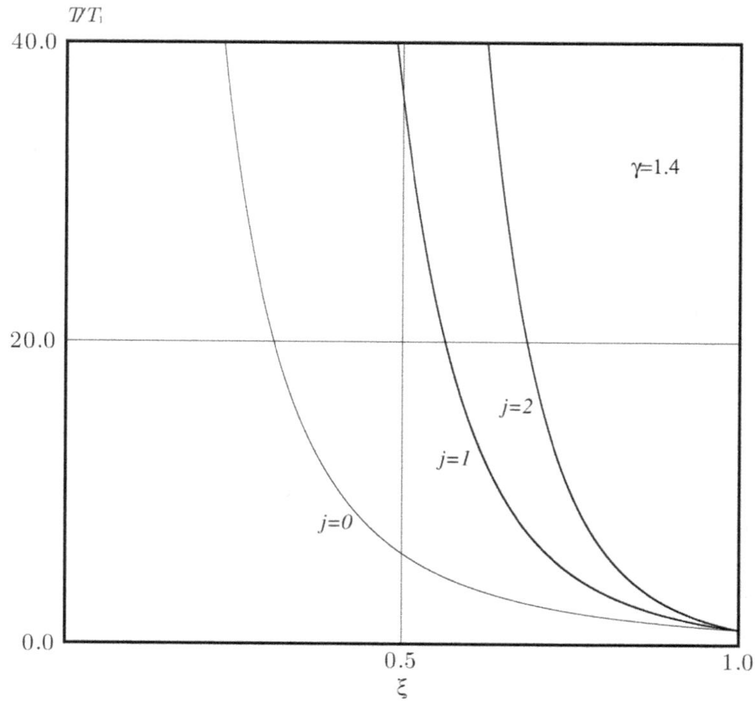

Figure 4.5 Temperature Distribution Behind the Shock Wave

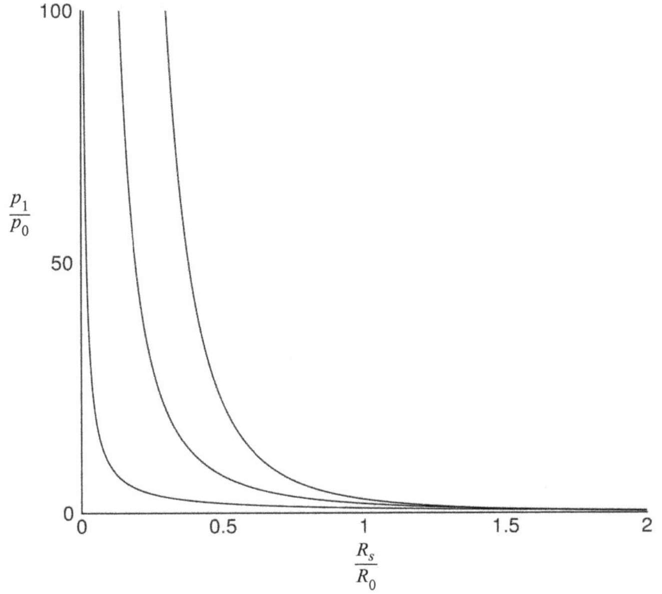

Figure 4.6 The Shock Pressure as a Function of Shock Radius

The density profile shown in Fig. 4.2 reveals that the mass is fairly concentrated at the shock front. With the density decreasing to a very low value toward the center of symmetry while the pressure remains constant (Fig. 4.3), the temperature increases behind the shock front to a very high level toward the center of symmetry. Thus the temperature of the core of the blast wave is extremely high (Fig. 4.5).

From the solutions for $\phi(\xi)$, $\psi(\xi)$, and $f(\xi)$, the energy integral can be evaluated. The decay of the shock pressure (Eq 4.3.23) and the shock trajectory (Eq 4.3.24) with the shock radius $\frac{R_s}{R_0}$ can readily be obtained. These are shown in Figs. 4.6 and 4.7.

4.4 Integrals of the Similarity Equations

The similarity equations given by Eqs 4.2.6–4.2.8 can also be integrated analytically.

Dividing Eq 4.2.6 by $\psi(\phi - \xi)$ and re-arranging gives

$$\frac{\psi'}{\psi} + \frac{(\phi - \xi)'}{(\phi - \xi)} + \frac{j}{\xi} + \frac{j+1}{(\phi - \xi)} = 0. \tag{4.4.1}$$

Similarly, dividing Eq 4.2.8 by $f(\phi - \xi)$ yields

$$\frac{f'}{f} + \frac{\gamma(\phi - \xi)'}{(\phi - \xi)} + \frac{\gamma j}{\xi} + \frac{2\theta + 2(j+1)}{(\phi - \xi)} = 0. \tag{4.4.2}$$

Eliminating $\frac{1}{(\phi - \xi)}$ and adding the two equations give

$$[2\theta + \gamma(j+1)]\frac{\psi'}{\psi} - (j+1)\frac{f'}{f} + \frac{2\theta(\xi - \phi)'}{(\xi - \phi)} + \frac{2\theta j}{\xi} = 0,$$

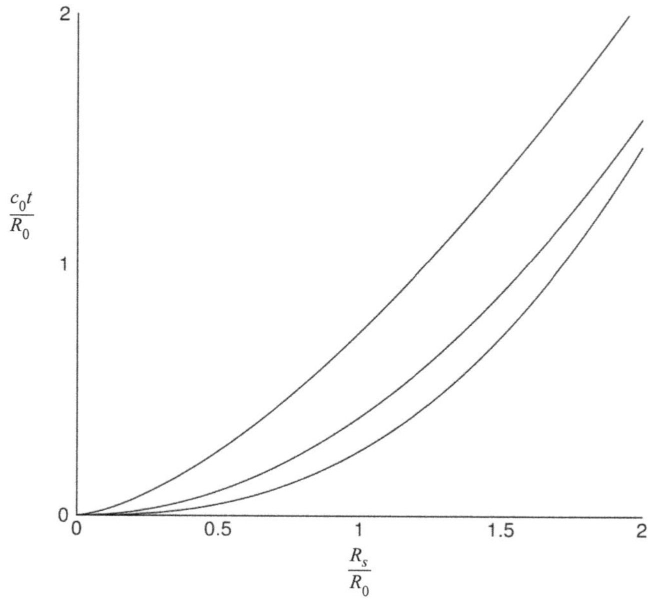

Figure 4.7 The Shock Trajectory as a Function of Shock Radius

which integrates to yield the so-called first integral, that is,

$$\frac{\psi(\xi-\phi)^{\frac{2\theta}{\gamma(j+1)+2\theta}}\xi^{\frac{2\theta j}{\gamma(j+1)+2\theta}}}{f^{\frac{j+1}{\gamma(j+1)+2\theta}}} = \text{constant}. \tag{4.4.3}$$

The constant of integration can be evaluated using the boundary condition at the shock front $\xi = 1$, that is,

$$\left[\frac{(\gamma+1)^{\gamma+1}}{2(\gamma-1)^{\gamma}}\right]^{j+1}, \tag{4.4.4}$$

where the value of $\theta = -\frac{j+1}{2}$ has been used.

A second integral can be obtained by multiplying Eq 4.2.6 by $\frac{\phi^2}{2}$, Eq 4.2.7 by $\phi\psi$, and Eq 4.2.8 by $\frac{1}{\gamma-1}$. Adding the resulting equations and re-arranging yields

$$(\phi-\xi)i' + (\phi-\xi)'i + \frac{j(\phi-\xi)i}{\xi} + \phi' f + f'\phi + \frac{jf\phi}{\xi} + (2\theta+j+1)i = 0, \tag{4.4.5}$$

where $i = \frac{f}{\gamma-1} + \frac{\psi\phi^2}{2}$. For the particular value of $\theta = -\frac{j+1}{2}$, the last term in the above equation vanishes and integrating Eq 4.4.5 then yields

$$\xi^j[(\phi-\xi)i + f\phi] = \text{constant}.$$

Using the strong shock condition (i.e., Eqs 4.2.14–4.2.16), the constant of integration is found to be zero and we obtain the second integral as

$$(\phi-\xi)\left(\frac{f}{\gamma-1} + \frac{\psi\phi^2}{2}\right) + f\phi = 0. \tag{4.4.6}$$

Unlike the first integral (i.e., Eq 4.4.3) which is valid for arbitrary values of θ, the second integral (Eq 4.4.6) requires that $\theta = -\left(\frac{j+1}{2}\right)$.

A third integral can also be obtained using Eq 4.2.12 for ϕ', that is,

$$\phi' = \frac{-\theta\phi(\phi-\xi) + \frac{f}{\psi}\left(2\theta + \frac{\gamma j \phi}{\xi}\right)}{(\phi-\xi)^2 - \frac{\gamma f}{\psi}}.$$

From the second integral (Eq 4.4.6), we can obtain $\frac{f}{\psi}$ in terms of ϕ and ξ as

$$\frac{f}{\psi} = -\left(\frac{\gamma-1}{2}\right)\frac{\phi^2(\phi-\xi)}{(\gamma\phi-\xi)}.$$

Substituting the above equation into Eq 4.2.12 and using the value of $\theta = -\left(\frac{j+1}{2}\right)$, we obtain

$$\phi' = -\frac{\phi}{\xi}\frac{\left[\gamma j(\gamma-1)\phi^2 - \phi\xi[(j+1)(2\gamma-1)] + (j+1)\xi^2\right]}{[\gamma(\gamma+1)\phi^2 - 2(\gamma+1)\phi\xi + 2\xi^2]}.$$

Defining a new variable $\chi = \frac{\phi}{\xi}$, the above equation can be written as

$$-\frac{d\xi}{\xi} = d\chi\left(\frac{A_1}{\chi} + \frac{A_2}{\chi-\alpha_1} + \frac{A_3}{\chi-\alpha_2}\right), \tag{4.4.7}$$

where

$$A_1 = \frac{2}{j+3} = N,$$

$$A_2 = \frac{\gamma+1}{2+(\gamma-1)(j+1)} - A_1 - A_3,$$

$$A_3 = -\frac{\gamma-1}{j+1} + 2(\gamma-1),$$

$$\alpha_1 = \frac{j+3}{2+(j+1)(\gamma-1)},$$

$$\alpha_2 = \frac{1}{\gamma}.$$

Integrating Eq 4.4.7 yields

$$\frac{K_3}{\xi} = \left(\frac{\phi}{\xi}\right)^{A_1}\left(\frac{\phi}{\xi}-\alpha_1\right)^{A_2}\left(\frac{\phi}{\xi}-\alpha_2\right)^{A_3}. \tag{4.4.8}$$

The constant of integration K_3 can be evaluated using the strong shock condition (Eq 4.2.15) at $\xi = 1$ as

$$K_3 = \frac{2^{A_1}\left[\frac{(j+1)(\gamma-3)-2(\gamma-1)}{2+(j+1)(\gamma-1)}\right]^{A_2}\left(\frac{\gamma-1}{\gamma}\right)^{A_3}}{(\gamma+1)^{A_1+A_1+A_3}}. \tag{4.4.9}$$

The middle term in the numerator of the right hand side of Eq 4.4.9 is valid for the case $[(j+1)(\gamma-3)-2(\gamma-1)]>0$. For the case where $[(j+1)(\gamma-3)-2(\gamma-1)]<0$, Eq 4.4.8 should be written as

$$\frac{K_3}{\xi}=\left(\frac{\phi}{\xi}\right)^{A_1}\left(\alpha_1-\frac{\phi}{\xi}\right)^{A_2}\left(\frac{\phi}{\xi}-\alpha_2\right)^{A_3}, \tag{4.4.10}$$

and K_3 now takes the form

$$K_3=\frac{2^{A_1}\left[\frac{2(\gamma-1)-(j+1)(\gamma-3)}{2+(j+1)(\gamma-1)}\right]^{A_2}\left(\frac{\gamma-1}{\gamma}\right)^{A_3}}{(\gamma+1)^{A_1+A_1+A_3}}. \tag{4.4.11}$$

For the planar ($j=0$), cylindrical ($j=1$), and spherical cases ($j=2$), the term $(j+1)(\gamma-3)-2(\gamma-1)$ has the values $-(\gamma+1)$, -4, $\gamma-7$, respectively. Since $\gamma>0$, we see that for planar and cylindrical blasts, $[(j+1)(\gamma-3)-2(\gamma-1)]$ is always less than zero. Thus the form of the third integral given by Eq 4.4.10 is always valid. For the spherical geometry, Eq 4.4.10 is valid for $\gamma<7$. For most practical cases $\gamma<7$ and thus the third integral as given by Eq 4.4.10 is the form to be used in general for all three geometries.

4.5 Closed Form Solution for Blasts

From the three integrals given by Eqs 4.4.3, 4.4.6, and 4.4.10 the solutions for $\phi(\xi)$, $\psi(\xi)$, and $f(\xi)$ can be obtained. The simplest way to obtain the solution is to use the third integral to determine ξ in terms of $\frac{\phi}{\xi}$ first. Then $f(\xi)$ and $\psi(\xi)$ can be found from the first and second integrals. Using the value of $\theta=-\left(\frac{j+1}{2}\right)$, Eq 4.4.3 can be written in the following form

$$\left(\frac{\psi^{\gamma-1}}{f}\right)^{j+1}=\left[\frac{(\gamma+1)^{\gamma+1}}{2(\gamma-1)\gamma}\right]^{j+1}\xi^{j(j+1)}(\xi-\phi)^{j+1}, \tag{4.5.1}$$

From the second integral, we obtain,

$$\frac{f}{\psi}=\left(\frac{\gamma-1}{2}\right)\frac{\phi^2(\xi-\phi)}{(\gamma\phi-\xi)}. \tag{4.5.2}$$

Solving for ψ by eliminating f from the above equations yields

$$\psi^{(\gamma-2)(j+1)}=\left[\frac{(\gamma+1)^{\gamma+1}\phi^2}{4(\gamma-1)^{\gamma-1}(\gamma\phi-\xi)}\right]^{j+1}(\xi-\phi)^{2(j+1)}\xi^{j(j+1)}. \tag{4.5.3}$$

Re-arranging, the above equation gives

$$\psi=\left[\frac{(\gamma+1)^{\gamma+1}\phi^2}{4(\gamma-1)^{\gamma-1}(\gamma\phi-\xi)}\right]^{\frac{1}{\gamma-2}}(\xi-\phi)^{\frac{2}{\gamma-2}}\xi^{\frac{j}{\gamma-2}}. \tag{4.5.4}$$

Substituting the above equation for ψ into Eq 4.5.2, f can be obtained as

$$f = \left(\frac{\gamma - 1}{2}\right)\left[\frac{(\gamma+1)^{\gamma+1}}{4(\gamma-1)^{\gamma-1}}\right]^{\frac{1}{\gamma-2}}\left[\frac{\phi^2}{\gamma\phi - \xi}\right]^{\frac{\gamma-1}{\gamma-2}} \xi^{\frac{j}{\gamma-2}}(\xi - \phi)^{\frac{2}{\gamma-2}}. \tag{4.5.5}$$

The third integral gives ϕ as a function of ξ. Thus solutions for $\phi(\xi)$, $\psi(\xi)$, and $f(\xi)$ are given by Eqs 4.4.10, 4.5.4 and 4.5.5, respectively. The solutions can be written in a more convenient form by defining the following parameters

$$\left.\begin{aligned}
\beta_1 &= \frac{1}{\phi(1)}\left(\frac{\phi}{\xi}\right)\\
\beta_2 &= \frac{1}{\phi(1)-1}\left(\frac{\phi}{\xi} - 1\right)\\
\beta_3 &= \frac{1}{\gamma\phi(1)-1}\left(\frac{\gamma\phi}{\xi} - 1\right)\\
\beta_4 &= \frac{1}{\frac{\phi(1)}{\alpha_1} - 1}\left(\frac{\phi}{\alpha_1\xi} - 1\right)
\end{aligned}\right\} \tag{4.5.6}$$

where $\phi(1) = \frac{2}{\gamma+1}$ is the value of ϕ at the shock front. Note that all the β's are unity at the shock front $\xi = 1$.

Using Eq 4.5.6, the third integral can conveniently be expressed as

$$\xi = \beta_1^{-\frac{2}{j+3}} \beta_3^{\frac{\gamma-1}{(j+1)+2(\gamma-1)}} \beta_4^{-A_2}, \tag{4.5.7}$$

where A_2 is given previously by Eq 4.4.7. In terms of the β's, the solutions for $\psi(\xi)$ and $f(\xi)$ are obtained as

$$\frac{\psi(\xi)}{\psi(1)} = \beta_2^{\frac{2}{\gamma-2}} \beta_3^{\frac{j+1}{(j+1)+2(\gamma-1)}} \beta_4^{\frac{-A_2(j+3)}{\gamma-2}}, \tag{4.5.8}$$

$$\frac{f(\xi)}{f(1)} = \beta_1^2\beta_2\beta_3^{-1}\frac{\psi(\xi)}{\psi(1)}\xi^2, \tag{4.5.9}$$

where $\psi(\xi)$ and ξ are given by Eqs 4.5.8 and 4.5.7, respectively.

To obtain the temperature profile, we note that the sound speed can be written as

$$c^2 = \frac{\gamma p}{\rho} = \gamma RT.$$

Thus

$$\frac{T(\xi)}{T_s(1)} = \frac{f(\xi)}{f(1)}\frac{\psi(1)}{\psi(\xi)}, \tag{4.5.10}$$

where $T_s(1)$ is the temperature at the shock front. Solutions for $f(\xi)$ and $\psi(\xi)$ are given by Eqs 4.5.9 and 4.5.8, respectively.

Since the entropy of a fluid particle remains constant after crossing the shock, then $\frac{p}{\rho^\gamma}$ = constant. The Lagrangian coordinate of a particle can be expressed in terms of the pressure and density. Denoting the Lagrangian coordinate of a particle by $r_p(t)$ whose

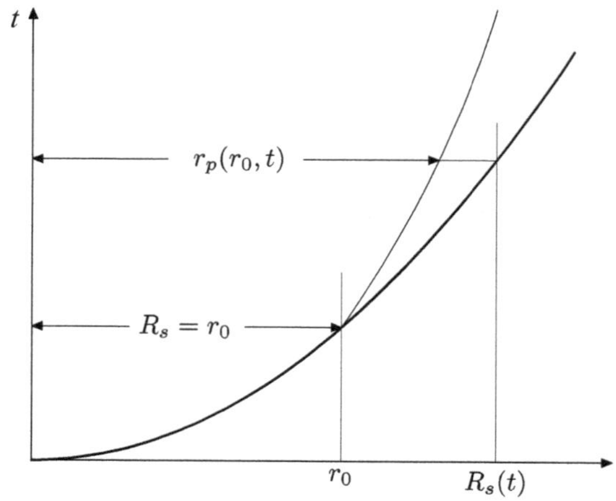

Figure 4.8 Lagrangian Coordinate of a Particle Whose Initial Position Is r_0

initial position is r_0, as shown in Fig. 4.8, we can write

$$\frac{p(r_p, R_s)}{\rho^\gamma(r_p, R_s)} = \frac{p(r_0, R_s = r_0)}{\rho^\gamma(r_0, R_s = r_0)}.$$

Since $f(\xi) = \frac{p(r,R_s)}{\rho_0 \dot{R}_s^2}$ and $\psi(\xi) = \frac{\rho(r,R_s)}{\rho_0}$, we can write

$$\frac{\rho_0 \dot{R}_s^2 f(\xi)}{\rho_0^\gamma \psi^\gamma(\xi)} = \frac{\rho_0 \dot{r}_0^2 f(1)}{\rho_0^\gamma \psi^\gamma(1)},$$

hence

$$\frac{f(\xi)}{\psi^\gamma(\xi)} = \frac{f(1)}{\psi^\gamma(1)} \frac{\dot{r}_0^2}{\dot{R}_s^2}.$$

Note that $\dot{r}_0^2$ is the shock velocity when the shock is at $R_s = r_0$. Using the energy integral (Eq 4.3.9), we obtain

$$\frac{\dot{r}_0^2}{\dot{R}_s^2} = \left(\frac{R_s}{r_0}\right)^{j+1},$$

hence

$$\frac{f(\xi)}{\psi^\gamma(\xi)} = \frac{f(1)}{\psi(1)} \left(\frac{R_s}{r}\right)^{j+1} \left(\frac{r}{r_0}\right)^{j+1},$$

and solving for $\frac{r}{r_0}$, we get

$$\frac{r}{r_0} = \left[\frac{f(\xi)}{f(1)} \left(\frac{\psi(1)}{\psi(\xi)}\right)^\gamma \xi^{j+1}\right]^{\frac{1}{j+1}}, \quad (4.5.11)$$

In the above equation r is the position of the fluid particle originally at r_0 when it crosses the shock. Solutions for $f(\xi)$ and $\psi(\xi)$ are given by Eqs 4.5.8 and 4.5.9.

To determine the profiles $\psi(\xi)$, $\phi(\xi)$, and $f(\xi)$ for a given geometry j and value of γ, a convenient procedure could be to first choose a value of $\frac{\phi}{\xi}$. The value of β_1, β_2, β_3, and β_4 are then evaluated (Eq 4.5.6). With the βs known, the profiles can then be computed from Eqs 4.5.8 and 4.5.9. The temperature profile is determined using Eq 4.5.10 and the Lagrangian coordinate from Eq 4.5.11.

Although a closed form solution can be obtained for the strong blast problem, the expressions are implicit and complicated. Thus, direct numerical integration of the set of similarity equations is more convenient in general.

4.6 Properties of the Constant Energy Solution

From the integrals and the closed form solutions obtained, the properties of the constant energy solution can be investigated. Consider first the behavior of the solution near the center of symmetry $\xi = 0$. For $\xi \ll 1$, we may write the solutions in the form of a power series in ξ, that is,

$$\phi(\xi) = \phi_0 \xi^{a_1} + \phi_1 \xi^{a_2} + \cdots$$

$$\psi(\xi) = \psi_0 \xi^{b_1} + \psi_1 \xi^{b_2} + \cdots$$

$$f(\xi) = \phi_0 \xi^{c_1} + f_1 \xi^{c_2} + \cdots$$

Substituting the expression for $\phi(\xi)$ into the third integral (Eq 4.4.10), we find that $a_1 = 1$ and $\phi_0 = \frac{1}{\gamma}$ or $\frac{j+3}{2+(j+1)(\gamma-1)}$. The correct solution for ϕ_0 is $\frac{1}{\gamma}$ since the latter yields an unsatisfactory value for a_2 (a_2 must be greater than unity for ϕ' to be finite as $\xi \to 0$). Using $\phi_0 = \frac{1}{\gamma}$, a_2 is obtained as

$$a_2 = \frac{j+1}{\gamma - 1} + 3.$$

And the coefficient ϕ_1 is found to be

$$\phi_1 = \left[\left(\frac{1}{\gamma}\right)^{\frac{2}{j+3}} \left(\frac{j+3}{2+(j+1)(\gamma-1)} - \frac{1}{\gamma} \right)^{A_2} \frac{1}{K_3} \right]^{\frac{j+1+2(\gamma-1)}{\gamma-1}}, \tag{4.6.1}$$

where A_2 and K_3 are given by Eqs 4.4.7 and 4.4.9. Substituting the expansions for $\psi(\xi)$ and $f(\xi)$ into Eqs 4.5.4 and 4.5.5 yields

$$\psi_0 = \left[\frac{(\gamma+1)^{\gamma+1}}{4(\gamma-1)^{\gamma-3}\gamma^5 \phi_1} \right]^{\frac{1}{\gamma-2}}, \tag{4.6.2}$$

$$\left.\begin{aligned} b_1 &= \frac{j+1}{\gamma-1} \\ f_0 &= \frac{(\gamma-1)^2 \psi_0}{2\gamma^4 \phi_1} \\ c_1 &= 0 \\ c_2 &= b_1 + 2 \end{aligned}\right\} \tag{4.6.3}$$

Thus the solution near the center of symmetry ($\xi = 0$) can be written as

$$\left.\begin{aligned} \phi(\xi) &= \frac{1}{\gamma} + \phi_1 \xi^{\frac{j+1}{\gamma-1}+3} + \cdots \\ \psi(\xi) &= \left[\frac{(\gamma+1)^{\gamma+1}}{4(\gamma-1)^{\gamma-3}\gamma^5\phi_1}\right]^{\frac{1}{\gamma-2}} \xi^{\frac{j+1}{\gamma-1}} + \cdots \\ f(\xi) &= \frac{(\gamma-1)^2\psi_0}{2\gamma^4\phi_0} + f_1 \xi^{\frac{j+1}{\gamma-1}+2} + \cdots \end{aligned}\right\} \tag{4.6.4}$$

where ϕ_1 and ψ_1 are given by Eqs 4.6.1 and 4.6.2. For the particular case of spherical blast ($j = 2$) and for $\gamma = 1.4$, Eq 4.6.4 becomes

$$\begin{aligned} \phi(\xi) &= \frac{5}{7}\xi + \phi_1 \xi^{\frac{21}{2}} + \cdots \\ \psi(\xi) &= \psi_0 \xi^{\frac{15}{2}} + \cdots \\ f(\xi) &= f_0 + f_1 \xi^{\frac{19}{2}} + \cdots \end{aligned} \tag{4.6.5}$$

Note that the exponent of ξ is a fairly large number, hence the solution given by Eq 4.6.5 is valid for values of ξ even quite far away from the center. The exponent of the density profile is also fairly large indicating that the mass behind the blast wave is concentrated near the shock front (particularly when $\gamma \simeq 1$). For example, for $j = 2$, $\gamma = 1.4$, $\psi(\xi) \to \xi^{\frac{j+1}{\gamma-1}} = \xi^{30}$. Thus, the blast resembles a thin spherical shell where most of the shocked gas is concentrated and very little mass is found away from the shock front.

In the above consideration, we have considered the case where $(j+1)(\gamma-3) - 2(\gamma-1) < 0$ and so the third integral is of the form given by Eq 4.4.10. For $[(j+1)(\gamma-3) - 2(\gamma-1)] > 0$, the third integral is given by Eq 4.4.8. It is found that there is a value $\xi = \xi_0$ when $\psi = 0$. Thus a vacuum "hole" exists. Near the vacuum edge ξ_0, the solution can be written as

$$\left.\begin{aligned} \phi(\xi) &= \xi_0 + \phi_1^*(\xi - \xi_0) + \cdots \\ \psi(\xi) &= \psi_0^*(\xi - \xi_0)^{\frac{2}{\gamma-2}} + \cdots \\ f(\xi) &= f_0^*(\xi - \xi_0)^{\frac{2}{\gamma-2}} + \cdots \\ \text{with} \qquad \xi_0 &= \frac{C\left(\frac{\gamma-1}{\gamma}\right)^{\frac{\gamma-1}{2\gamma+1}}}{\left(1 - \frac{5}{3\gamma-1}\right)^{A_2}} \end{aligned}\right\} \tag{4.6.6}$$

The coefficients ϕ_1^*, ψ_0^*, and t_0^* can be determined by substituting the expansions (Eq 4.6.6) into the third integral and using the solutions for $\psi(\xi)$ and $f(\xi)$. For large values of γ, the exponent of $(\xi - \xi_0)$ in the density profile is small (i.e., $\frac{2}{\gamma-2} \ll 1$ when

$\gamma \gg 1$). Thus the density increases sharply from zero for $\xi > \xi_0$. The vacuum interface is essentially a density interface. In general, the values of $\gamma \simeq 1$, and thus the form for the third integral given by Eq 4.4.8 applies.

For the particular case of spherical blast ($j = 2$) and $\gamma = 7$, the solution is particularly simple. From Eq 4.4.11, we note that $K_3 = 0$, and thus the solution for ϕ (Eq 4.4.10) gives a linear velocity profile

$$\frac{\phi}{\xi} = \frac{1}{4}. \tag{4.6.7}$$

The solutions for $\psi(\xi)$ and $f(\xi)$ (Eqs 4.5.4, 4.5.5) reduce to

$$\psi(\xi) = \frac{4}{3}\xi, \tag{4.6.8}$$

$$f(\xi) = \frac{\xi^3}{4}. \tag{4.6.9}$$

Thus the particle velocity and density profiles are linear and the pressure profile is cubic with ξ. In general the profiles can readily be obtained by numerical integration of the similarity equations.

4.7 Variable Energy Blasts

For energy release that is a function of time, self-similar solution is still possible for strong shocks. The blast wave generated by an expanding piston is a practical example of a variable energy blast wave. The piston problem is analogous to the hypersonic flow over blunted nosed slender bodies.

Consider the flow generated by an expanding piston $R_p(t)$ and the flow is bounded by the shock and the piston, that is, $R_p \leq r \leq R_s$, or $\xi_p \leq \xi \leq 1$. The flow is governed by the self-similar equations (i.e., Eqs 4.2.11–4.2.13) and the boundary condition at the shock front $\xi = 1$ is given by Eqs 4.2.14–4.2.16. The value of θ can be determined from the energy integral, that is,

$$E_0(t) = \int_{R_p}^{R_s} \left(\frac{p}{\gamma - 1} + \frac{\rho u^2}{2} \right) k_j r^j \mathrm{d}r.$$

In terms of the similarity variables, the above can be written as

$$E_0(t) = \rho_0 k_j \dot{R}_s^2 R_s^{j+1} I. \tag{4.7.1}$$

where

$$I = \int_{\xi_p}^{1} \left(\frac{f}{\gamma - 1} + \frac{\psi\phi^2}{2} \right) \xi^j \mathrm{d}\xi.$$

We may express $E_0(t)$ in terms of the shock radius $R_s(t)$ and write

$$E_0(t) = B R_s^{\alpha}(t), \tag{4.7.2}$$

where $E_0(t)$ is expressed as a power law of $R_s(t)$. When $\alpha = 0$, we obtain the constant energy solution considered in previous sections. Using Eq 4.7.2, the energy integral (Eq 4.7.1) becomes

$$B = \rho_0 k_j C^2 R_s^{2\theta+j+1-\alpha} I, \tag{4.7.3}$$

where for self-similar solutions, $\theta =$ constant and $\dot{R}_s = CR_s^\theta$. Since the left hand side of Eq 4.7.3 is constant, the right hand side must not be explicitly dependent on time (or $R_s(t)$). Thus the exponent of R_s must vanish giving

$$\theta = -\frac{(j+1)-\alpha}{2}. \tag{4.7.4}$$

When $\alpha = 0$, we obtain $\theta = -\left(\frac{j+1}{2}\right)$ for the constant energy solution discussed previously.

For $\alpha = j+1$, Eq 4.7.4 gives $\theta = 0$ and $\dot{R}_s =$ constant and $R_s \sim t$, that is, constant velocity shock wave driven by a constant velocity piston. For $\alpha > j+1$ we note that $\theta > 0$ and $\dot{R}_s \to 0$ as $R_s \to 0$. Thus, the shock accelerates from zero velocity initially violating the self-similar requirement. Therefore, $0 \leq \alpha \leq j+1$ and the corresponding range for θ is

$$-\left(\frac{j+1}{2}\right) \leq \theta \leq 0. \tag{4.7.5}$$

Since $\theta = -\left(\frac{N-1}{N}\right)$ the range of values for the time exponent of the shock trajectory $R_s \sim t^N$ is

$$\frac{2}{j+3} \leq N \leq 1. \tag{4.7.6}$$

Once the value of α is specified, the value of θ can be obtained (Eq 4.7.4) and the self-similar equations (i.e., Eqs 4.2.11–4.2.13) can be integrated using the strong shock conditions (i.e., Eqs 4.2.14–4.2.16) at the shock front $\xi = 1$. The integration terminates at the piston surface $\xi = \xi_p$ where $\phi = \phi_p = \xi_p$ (particle velocity equals to the piston velocity at the piston face). Also, in order for the density to be finite at the piston surface, certain restrictions on the range of θ are imposed. From the first integral obtained previously, that is, Eq 4.4.3

$$\frac{\psi(\xi-\phi)^{\frac{2\theta}{2\theta+\gamma(j+1)}}\xi^{\frac{2\theta j}{2\theta+\gamma(j+1)}}}{f^{\frac{j+1}{2\theta+\gamma(j+1)}}} = \text{constant}, \tag{4.4.3}$$

we note that as $(\xi - \phi) \to 0$, the density $\psi \to (\xi-\phi)^{-\frac{2\theta}{2\theta+\gamma(j+1)}}$ if the left hand side of Eq 4.4.3 is to remain a constant. Furthermore, if ψ is to be finite as $(\xi - \phi) \to 0$, then

$$-\frac{2\theta}{2\theta+\gamma(j+1)} > 0. \tag{4.7.7}$$

Since $\theta = -\left(\frac{1-N}{N}\right)$, the above equation can alternately be written as

$$\frac{2(1-N)}{N\gamma(j+1)-2(1-N)} > 0, \tag{4.7.8}$$

where N is the time exponent of the shock trajectory $R_s \sim t^N$. Since $N < 1$, we require that $N\gamma(j+1) - 2(1-N) > 0$ or $N(2+\gamma(j+1)) - 2 > 0$. Thus

$$N > \frac{2}{2+\gamma(j+1)},$$

However, from Eq 4.7.4

$$N = \frac{2}{j+3-\alpha},$$

hence

$$\frac{2}{j+3-\alpha} > \frac{2}{2+\gamma(j+1)},$$

or $j+3-\alpha < 2+\gamma(j+1)$ which gives $\alpha > -(\gamma-1)(j+1)$. For a shock wave to exist at $t=0$, we see that $\alpha \leq j+1$. Thus

$$-(\gamma-1)(j+1) \leq \alpha \leq j+1. \tag{4.7.9}$$

If α is within the limits stated in Eq 4.7.9 above, then the density will be finite at the piston surface $(\phi - \xi) = 0$. Since $\alpha = 0$ corresponds to the constant energy blast, the range of values for α in practice is given by

$$0 \leq \alpha \leq j+1. \tag{4.7.10}$$

The shock trajectory is given by $R_s = At^N$, with $N = \frac{2}{j+3-\alpha}$, the energy integral (Eq 4.7.1) gives the value of A as

$$A = \left[\frac{B}{\rho_0 k_j I N^2}\right]^{\frac{1}{j+3-\alpha}}.$$

Defining an explosion length R_0 as

$$R_0 = \left(\frac{B}{p_0}\right)^{\frac{1}{j+1-\alpha}},$$

the shock trajectory can be obtained as

$$\frac{R_s}{R_0} = \left[\left(\frac{j+3-\alpha}{2}\right)^2 \frac{1}{\gamma k_j I}\right]^{\frac{1}{j+3-\alpha}} \left(\frac{c_0 t}{R_0}\right)^{\frac{2}{j+3-\alpha}}. \tag{4.7.11}$$

The Mach number square M_s^2 can be obtained from Eq 4.7.11 by differentiation as

$$M_s^2 = \frac{1}{\gamma k_j I} \frac{1}{\left(\frac{R_s}{R_0}\right)^{(j+1)-\alpha}}. \tag{4.7.12}$$

The shock pressure can also be obtained from Eq 4.7.12 as

$$\frac{p_1}{p_0} = \gamma M_s^2 \left(\frac{2}{\gamma+1}\right) = \left(\frac{2}{\gamma+1}\right)\left(\frac{1}{k_j I}\right)\frac{1}{\left(\frac{R_s}{R_0}\right)^{j+1-\alpha}}. \tag{4.7.13}$$

The above results reduce to the constant energy blast solution when $\alpha = 0$. The dimensionless integral I can be evaluated when the similarity equations are integrated for given value of α and γ.

Alternately we may include γ and k_j in the definition of the explosion length, that is,

$$R_0 = \left(\frac{B}{\gamma k_j p_0}\right)^{\frac{1}{j+1-\alpha}}.$$

In terms of this alternate definition, Eqs 4.7.11 and 4.7.12 get simplified to

$$\frac{R_s}{R_0} = \left[\left(\frac{j+3-\alpha}{2}\right)^2 \frac{1}{I}\right]^{\frac{1}{j+3-\alpha}} \left(\frac{c_0 t}{R_0}\right)^{\frac{2}{j+3-\alpha}},$$

and

$$M_s^2 = \frac{1}{I}\left(\frac{R_0}{R_s}\right)^{j+1-\alpha}.$$

5 Homentropic Explosions

5.1 Introduction

There is a class of explosion problems in which the explosion front propagates at a constant velocity. Thus the entropy change across the front is the same for every particle and, if there are no further entropy-generating mechanisms behind the front, the entire flow field behind the front will be isentropic. Flow with uniform entropy throughout is refered to as homentropic flow. For lack of a better name, we shall refer to this class of problems as "homentropic explosions."

For isentropic flow, there exists a relationship between the pressure and the density (i.e., the isentrope). Thus using the isentropic relationships, the number of dependent variables can be reduced, and accordingly, one of the three conservation equations can be eliminated.

The similarity equations for one-dimensional planar, cylindrical, and spherical symmetries derived previously are given by

$$(\phi - \xi)\psi' + \psi\phi' + \frac{j\phi\psi}{\xi} = 0, \tag{5.1.1}$$

$$(\phi - \xi)\phi' + \theta\phi + \frac{f'}{\psi} = 0, \tag{5.1.2}$$

$$(\phi - \xi)f' + 2\theta f + \gamma f\phi' + \frac{\gamma j\phi f}{\xi} = 0, \tag{5.1.3}$$

where $\psi = \frac{\rho}{\rho_0}$, $\phi = \frac{u}{\dot{R}_s}$, $f = \frac{p}{\rho_0 \dot{R}_s^2}$, $\xi = \frac{r}{R_s}$ and $\theta = \frac{R_s \ddot{R}_s}{\dot{R}_s^2}$, as discussed earlier. For a constant velocity shock front, $R_s \sim t$, hence $\theta = 0$ in the above equations. For homentropic flow, the sound speed is generally chosen as a dependent variable instead of p and ρ independently. The sound speed c is given by

$$c^2 = \frac{\gamma p}{\rho} = \frac{\gamma f}{\psi}\dot{R}_s^2, \tag{5.1.4}$$

and defining a new variable β as

$$\beta^2 = \frac{c^2}{\dot{R}_s^2} = \frac{\gamma f}{\psi}, \tag{5.1.5}$$

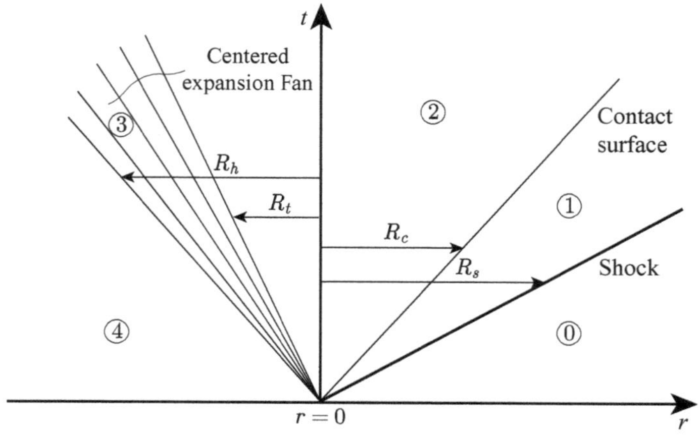

Figure 5.1 Shock Wave, Contact Surface, and Expansion Fan in the Shock Tube Problem

Eqs 5.1.1–5.1.3 can be reduced to

$$\phi' = \frac{j\phi\beta^2}{\xi[(\phi - \xi)^2 - \beta^2]}, \tag{5.1.6}$$

$$\beta' = \frac{-(\gamma - 1)(\phi' - \xi)j\phi\beta}{2\xi[(\phi - \xi)^2 - \beta^2]}, \tag{5.1.7}$$

for the two dependent variables β, ϕ. The boundary conditions at the shock front $\xi = 1$ are given by the Rankine–Hugoniot relationships,

$$\phi(1) = \frac{2}{\gamma + 1}(1 - \eta), \tag{5.1.8}$$

$$\beta(1) = \sqrt{\frac{\gamma f(1)}{\psi(1)}} = \left(\frac{1}{\gamma + 1}\right)[(2\gamma - (\gamma - 1)\eta)(\gamma - 1 + 2\eta)]^{\frac{1}{2}}, \tag{5.1.9}$$

where $\eta = \frac{1}{M_s^2}$ and, $M_s = \frac{\dot{R}_s}{c_0}$ where $c_0^2 = \frac{\gamma p_0}{\rho_0}$ is the sound speed in the medium ahead of the shock. Given a shock speed, Eqs 5.1.8 and 5.1.9 give $\phi(1)$ and $\beta(1)$ at the front $\xi = 1$. The similarity equations (i.e., Eqs 5.1.6 and 5.1.7) can be integrated to some rear boundary that depends on the particular explosion problem.

5.2 The Shock Tube Problem

A simple and useful example of a planar explosion is the shock tube problem. Consider a high-pressure gas to occupy the negative half space $-\infty \leq r \leq 0$ and a low-pressure gas in the positive half space $0 \leq r \leq \infty$. A diaphragm separates the two regions initially. At $t > 0$, the diaphragm is suddenly removed and a shock wave propagates into the low-pressure region $r > 0$, and a centered expansion fan propagates into the high-pressure region $r < 0$. The wave diagram of the explosion is given in Fig. 5.1.

Across the shock, $0 \to 1$, the state $\phi(1)$ and $\beta(1)$ are given by Eqs 5.1.8 and 5.1.9 but the shock strength η (or M_s) is as yet unknown. The contact surface separates the shocked gas from the expanded gas from region 4. Thus the contact surface acts as a piston driving the shock front. The pressure and particle velocity are uniform across the contact surface, thus $\phi(2) = \phi(1)$ and $f(2) = f(1)$ (where $f = \frac{p}{\rho_0 \dot{R}_s^2}$ is the dimensionless pressure). Region 3 is a non-uniform region of unsteady isentropic flow and this region is governed by Eqs 5.1.6 and 5.1.7. The head of the expansion fan $r = R_h$ propagates at the sound speed c_4 of the high-pressure region, that is, $\dot{R}_h = c_4$.

For planar geometry where $j = 0$, Eqs 5.1.6 and 5.1.7 either give $\phi' = 0,\ \beta' = 0$ which implies that either the flow is uniform or $(\phi - \xi)^2 - \beta^2 = 0$ and ϕ' and β' can then be finite. The non-steady flow in region 3 corresponds to $(\phi - \xi)^2 - \beta^2 = 0$, thus $(\phi - \xi) = \pm\beta$. Dividing Eq 5.1.6 by 5.1.7 gives

$$\frac{\phi'}{\beta'} = -\frac{2}{\gamma - 1}\frac{\beta}{\phi - \xi}.$$

Thus $\frac{\phi'}{\beta'} = \mp\frac{2}{\gamma-1}$, which integrates to yield

$$\phi = \mp\frac{2}{\gamma - 1}\beta + \text{constant}.$$

Since $\mp\beta = -(\phi - \xi)$, the above equation gives

$$\phi(\xi) = \frac{2}{\gamma + 1}\xi + \text{constant}, \tag{5.2.1}$$

$$\beta(\xi) = \mp\left(\frac{\gamma - 1}{\gamma + 1}\right)\xi + \text{constant}. \tag{5.2.2}$$

We take the negative sign in Eq 5.2.2 since β decreases as ξ increases. At the head of the expansion fan, $\xi = \xi_h$, $\phi(\xi_h) = 0$ and $\beta(\xi_h)$ can be obtained from $(\phi - \xi)^2 - \beta^2 = 0$ where at $\xi = \xi_h$, $\phi(\xi_h) = 0$ and hence $\beta(\xi_h) = -\xi_h$. The solution for the non-steady expansion region is thus

$$\phi(\xi) = \frac{2}{\gamma + 1}(\xi - \xi_h), \tag{5.2.3}$$

$$\beta(\xi) = -\left(\frac{\gamma - 1}{\gamma + 1}\right)(\xi - \xi_h) - \xi_h, \tag{5.2.4}$$

where

$$\beta(\xi_h) = -\xi_h = \frac{c_4}{\dot{R}_s} = \left(\frac{p_4}{p_0}\frac{\rho_0}{\rho_4}\eta\right)^{\frac{1}{2}}. \tag{5.2.5}$$

Since $\beta = \left(\frac{\gamma f}{\psi}\right)^{\frac{1}{2}}$ and the isentropic condition give $\frac{f}{\psi^\gamma} = \text{constant}$, we have

$$f(\xi) = K\beta(\xi)^{\frac{2\gamma}{\gamma-1}},$$

where K is some constant. Using the above equation, we can write

$$\frac{f(\xi)}{f(\xi_h)} = \frac{\beta(\xi)^{\frac{2\gamma}{\gamma-1}}}{\beta(\xi_h)^{\frac{2\gamma}{\gamma-1}}}.$$

And since $\beta(\xi_h) = -\xi_h$, we get

$$f(\xi) = f(\xi_h)\left(\frac{\beta(\xi)}{-\xi_h}\right)^{\frac{2\gamma}{\gamma-1}}. \tag{5.2.6}$$

Further, using Eq 5.2.4 for $\beta(\xi)$, we get an expression for $f(\xi)$ as

$$f(\xi) = f(\xi_h)\left(1 + \frac{\gamma-1}{\gamma+1}\frac{(\xi-\xi_h)}{\xi_h}\right)^{\frac{2\gamma}{\gamma-1}}. \tag{5.2.7}$$

The pressure in region 4 is p_4. Thus

$$f(\xi_h) = \frac{p_4}{\rho_0\dot{R}_s^2} = \frac{p_4}{p_0}\frac{\gamma p_0}{\gamma\rho_0\dot{R}_s^2} = \frac{p_4}{p_0}\frac{\eta}{\gamma}. \tag{5.2.8}$$

Since the pressure is uniform across the contact surface, $p_2 = p_1$, where p_2 is the pressure at the tail of the expansion fan and p_1 is the pressure behind the shock. Thus $f(\xi_t) = f(1)$ and $f(1)$ from the Rankine–Hugoniot equation is

$$f(1) = \frac{2}{\gamma+1}\left(1 - \frac{\gamma-1}{2\gamma}\eta\right). \tag{5.2.9}$$

Thus Eq 5.2.7 gives

$$f(\xi_t) = f(\xi_h)\left(1 + \frac{\gamma-1}{\gamma+1}\frac{(\xi-\xi_h)}{\xi_h}\right)^{\frac{2\gamma}{\gamma-1}} = f(1), \tag{5.2.10}$$

Eq 5.2.3 gives

$$\phi(\xi_t) = \frac{2}{\gamma+1}(\xi_t - \xi_h), \tag{5.2.11}$$

and since the particle velocity is uniform across the contact surface, $\phi(\xi_t) = \phi(1)$, where $\phi(1)$ is given by Eq 5.1.8 as

$$\phi(1) = \frac{2}{\gamma+1}(1-\eta).$$

Thus from Eq 5.2.11, we get

$$\phi(\xi_t) = \frac{2}{\gamma+1}\xi_h\left(\frac{\xi_t}{\xi_h} - 1\right) = \frac{2}{\gamma+1}(1-\eta). \tag{5.2.12}$$

Using Eq 5.2.5 for ξ_h, we can obtain from Eq 5.2.12

$$\left(\frac{\xi_t}{\xi_h} - 1\right) = -(1-\eta)\left(\frac{p_4}{p_0}\frac{\rho_0}{\rho_4}\eta\right)^{-\frac{1}{2}}. \tag{5.2.13}$$

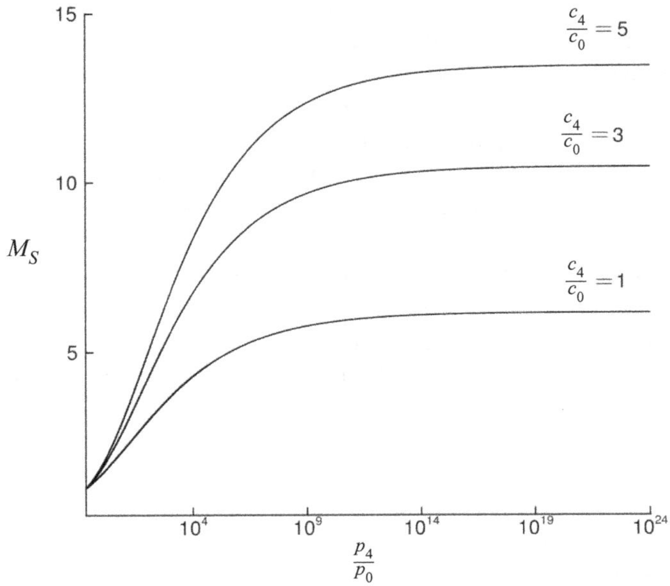

Figure 5.2 The Shock Mach Number as a Function of $\frac{p_4}{p_0}$

Using Eq 5.2.8 for $f(\xi_h)$, Eq 5.2.9 for $f(1)$, Eq 5.2.13 for $\left(\frac{\xi_l}{\xi_h} - 1\right)$, Eq 5.2.10 becomes

$$\left(\frac{p_4}{p_0}\right)\left(\frac{\eta}{\gamma}\right)\left[1 - \left(\frac{\gamma - 1}{\gamma + 1}\right)(1 - \eta)\left(\frac{p_4}{p_0}\frac{\rho_0}{\rho_4}\eta\right)^{-\frac{1}{2}}\right]^{\frac{2\gamma}{\gamma-1}} = \frac{2}{\gamma + 1}\left(1 - \frac{\gamma - 1}{2\gamma}\eta\right). \tag{5.2.14}$$

Since the sound speed $c^2 = \left(\frac{\gamma p}{\rho}\right)^{\frac{1}{2}}$, we can write $\left(\frac{p_4}{p_0}\frac{\rho_0}{\rho_4}\right)$ as $\left(\frac{c_4}{c_0}\right)$, and Eq 5.2.14 can alternately be written as

$$\left(\frac{p_4}{p_0}\right)\left(\frac{\eta}{\gamma}\right)\left[1 - \left(\frac{\gamma - 1}{\gamma + 1}\right)(1 - \eta)\left(\frac{c_4}{c_0}\eta\right)^{-\frac{1}{2}}\right]^{\frac{2\gamma}{\gamma-1}} = \frac{2}{\gamma + 1}\left(1 - \frac{\gamma - 1}{2\gamma}\eta\right). \tag{5.2.15}$$

Given initial values for $\frac{p_4}{p_0}$ and $\frac{c_4}{c_0}$, the shock strength $\eta = \frac{1}{M_s^2} = \left(\frac{c_0}{R_s}\right)^2$ can be obtained from Eq 5.2.15. Figure 5.2 shows the shock Mach number M_s, as a function of $\frac{p_4}{p_0}$ for various values of $\frac{c_4}{c_0}$ and a value of $\gamma = 1.4$. Note that for a given value of $\frac{c_4}{c_0}$, the shock Mach number M_s approaches a limiting value as $\frac{p_4}{p_0} \to \infty$.

5.3 Propagation of Chapman–Jouguet Detonations

Chapman–Jouguet detonations propagate at constant velocity once initiated. However, their existence depends on whether a solution for the non-steady expansion of the detonation products can be found that is compatible to the constant velocity Chapman–Jouguet detonation front. We shall discuss the dynamics of the detonation products behind a Chapman–Jouguet detonation for various geometries.

Consider first the planar geometry. We assume an explosion to occur in the positive half space $r \geq 0$ and the explosive is bounded by a solid wall at $r = 0$. At time $t = 0$, a Chapman–Jouguet detonation is initiated at $r = 0$ and propagates subsequently at a constant velocity. A centered rarefaction fan exists behind the detonation bringing the detonation products to rest to satisfy the boundary condition of zero velocity at the wall. The equations for the non-steady flow of the detonation products are given by Eqs 5.1.6 and 5.1.7 and for the planar geometry where $j = 0$, two possible solutions exist either $\phi' = \beta' = 0$ or ϕ' and β' finite but $(\phi - \xi)^2 - \beta^2 = 0$. The latter solution applies and we obtain the solution for $\phi(\xi)$ and $\beta(\xi)$ given by Eqs 5.2.1 and 5.2.2, that is,

$$\phi(\xi) = \frac{2}{\gamma + 1}\xi + \text{constant},$$

$$\beta(\xi) = \frac{\gamma - 1}{\gamma + 1}\xi + \text{constant}.$$

We take the positive sign in Eq 5.2.2 since β decreases with decreasing $\xi = \frac{r}{R_s}$ and $\dot{R}_s = D$ is the Chapman–Jouguet detonation. The constants of integration in Eqs 5.2.1 and 5.2.2 can be evaluated from the conditions at the detonation front $\xi = 1$, $\phi(\xi) = \phi(1)$, and $\beta(\xi) = \beta(1)$. For a strong Chapman–Jouguet detonation,

$$\phi(1) = \frac{1}{\gamma + 1}, \tag{5.3.1}$$

$$\beta(1) = \frac{\gamma}{\gamma + 1}, \tag{5.3.2}$$

where in Eqs 5.3.1 and 5.3.2, we have assumed the Chapman–Jouguet detonation Mach number to be sufficiently high, that is, $\frac{1}{M_s^2} \ll 1$. Note that Eqs 5.3.1 and 5.3.2 satisfy the Chapman–Jouguet condition that $D = u_1 + c_1$ or $\phi(1) + \beta(1) = 1$.

Using the boundary conditions at $\xi = 1$ to evaluate the constants, the solutions given by Eqs 5.2.1 and 5.2.2 become

$$\phi(\xi) = \frac{2}{\gamma + 1}\xi - \frac{1}{\gamma + 1}, \tag{5.3.3}$$

$$\beta(\xi) = \frac{\gamma - 1}{\gamma + 1}\xi + \frac{1}{\gamma + 1}. \tag{5.3.4}$$

Thus, the velocity decreases linearly behind the detonation front and $\phi = 0$ occurs when $\xi = \frac{1}{2}$. The sound speed $\beta(\xi)$ also decreases linearly with ξ and at $\xi = \frac{1}{2}$, $\beta = \frac{1}{2}$. The pressure and density profiles can be obtained from Eq 5.3.4 using the isentropic relationships. The dynamics of the combustion products behind a detonation wave was first investigated by Taylor (1950a) and Zel'dovich (1942). The expansion fan behind the detonation is generally referred to as the "Taylor wave."

For diverging cylindrical and spherical detonations, $j \neq 0$ and Eqs 5.1.6 and 5.1.7 have to be integrated numerically starting with the boundary conditions at $\xi = 1$, given by Eqs 5.3.1 and 5.3.2. However, for Chapman–Jouguet detonations, $\phi(1)+\beta(1)=1$. Since at the front $\xi=1$, $(\phi-\xi)^2-\beta^2=0$, ϕ' and β' become infinite. The presence of the singularity raises doubts about the existence of diverging spherical Chapman–Jouguet detonation waves. However, the view adopted by Taylor (1950a) is that the singularity of an infinite expansion gradient is no worse than the assumption of an infinite compression gradient for a shock or a detonation front. In practice, diverging spherical detonations require a relatively large initiation energy. Thus at small radius, the detonation is highly overdriven and decays asymptotically to a Chapman–Jouguet detonation at large radius. Thus, the model of a constant velocity Chapman–Jouguet detonation originating at $r=0$ initially is not realized experimentally and serves only as an approximation.

To integrate numerically Eqs 5.1.6 and 5.1.7, we must obtain an analytical solution in the vicinity of the front $\xi=1$. Near the Chapman–Jouguet front, we may write $\phi(\xi)$ and $\beta(\xi)$ in a series in $(1-\xi)$, that is,

$$\phi(\xi)=a_0+a_1(1-\xi)^x+\cdots \tag{5.3.5}$$

$$\beta(\xi)=b_0+b_1(1-\xi)^y+\cdots \tag{5.3.6}$$

Using the boundary conditions at $\xi=1$, that is, Eqs 5.3.1 and 5.3.2, we obtain

$$a_0=\phi(1)=\frac{1}{\gamma+1}, \tag{5.3.7}$$

$$b_0=\beta(1)=\frac{\gamma}{\gamma+1}. \tag{5.3.8}$$

For $\phi(\xi)$ and $\beta(\xi)$ to be finite as $\xi\to1$, the exponents x and y must be positive. From Eqs 5.3.5 and 5.3.6, we write

$$\phi'=a_1x(1-\xi)^{x-1}+\cdots$$

$$\beta'=b_1y(1-\xi)^{y-1}+\cdots$$

and since $\phi'\to\infty$ and $\beta'\to\infty$ as $\xi\to1$, the exponents x and y must be less than unity, that is, $0\le x\le1$, $0\le y\le1$. Substituting the expansion for $\phi(\xi)$ and $\beta(\xi)$ into Eqs 5.1.6 and 5.1.7, we obtain

$$\begin{aligned}2xa_1^2b_0(1-\xi)^{2x-1}&+2a_1xb_0b_1(1-\xi)^{x+y+1}\\&-j\left[a_0b_0^2+a_1b_0^2(1-\xi)^x+\cdots\right]=0,\\2yb_0b_1^2(1-\xi)^{2y-1}&+2yb_0b_1a_1(1-\xi)^{x+y+1}\\&-\frac{j(\gamma-1)}{2}a_0b_0^2+\cdots=0.\end{aligned}$$

Examining the first of the above equation, we note that j, a_0, b_0 are all finite. Thus we must have either $2x-1=0$ or $x+y-1=0$. From the second of the above equations we note that $2y-1$ or $x+y-1$ must be zero. Both requirements are satisfied if

$x = y = \frac{1}{2}$. Putting $x = y = \frac{1}{2}$ in the equations give

$$a_1 = \pm\sqrt{\frac{2j\phi(1)\beta(1)}{\gamma+1}},$$

$$b_1 = \pm\frac{\gamma-1}{2}\sqrt{\frac{2j\phi(1)\beta(1)}{\gamma+1}}.$$

Hence in the vicinity of the detonation front, the solution can be written as

$$\phi(\xi) = \phi(1) \pm \sqrt{\frac{2j\phi(1)\beta(1)}{\gamma+1}}(1-\xi)^{\frac{1}{2}} + \cdots \tag{5.3.9}$$

$$\beta(\xi) = \beta(1) \pm \frac{\gamma-1}{2}\sqrt{\frac{2j\phi(1)\beta(1)}{\gamma+1}}(1-\xi)^{\frac{1}{2}} + \cdots \tag{5.3.10}$$

where $\phi(1)$ and $\beta(1)$ are given by Eqs 5.3.1 and 5.3.2. In Eqs 5.3.9 and 5.3.10, the $\pm$ sign denote two possible solutions. The $+$ sign corresponds to the compression solution where both $\phi(\xi)$ and $\beta(\xi)$ increase further behind the detonation front. The compression solution requires a constant velocity piston to follow behind the detonation. The integration of Eqs 5.1.6 and 5.1.7 terminates at the piston surface $\xi = \xi_p$ where $(\phi - \xi) = 0$. The negative sign in Eqs 5.3.9 and 5.3.10 gives the expansion solution where $\phi \to 0$ and $\xi \to 0$ are at the center of symmetry. From the solution for $\beta(\xi)$, the pressure and density profiles behind the detonation can be obtained using the isentropic relationships.

Using Eqs 5.3.9 and 5.3.10 to obtain values for $\phi(\xi)$ and $\beta(\xi)$ near the detonation front, Eqs 5.1.6–5.1.7 are integrated numerically and the pressure and particle velocity profiles are shown in Figs. 5.3 and 5.4 for three geometries of planar, cylindrical, and spherical detonations. The Chapman–Jouguet boundary conditions at the front $\xi = 1$ are given by Eqs 5.3.1 and 5.3.2 and we choose a typical value of $\gamma = 1.2$.

5.4 Piston Driven Explosion

We consider a constant velocity shock or detonation driven by a steadily expanding piston. For the planar geometry, the flow behind the shock is uniform from the shock front to the piston surface. The shock strength is such that the particle velocity behind it is equal to the piston velocity. For cylindrical or spherical geometries, the flow behind the shock is non-uniform due to the area divergence between the piston and the shock.

For a low velocity expanding piston, the shock is weak and can be approximated by an acoustic wave. Thus we can use the acoustic equation for the velocity potential instead of the non-linear Euler equations. We can also seek a self-similar solution; however, the linear acoustic equation can be solved exactly. The solution for the acoustic radiation from an expanding piston was given previously in Section 1.5 of Chapter 1. For an arbitrary piston velocity where the shock strength is finite, Eqs 5.1.6 and 5.1.7 have to be integrated. Given a piston velocity, the shock speed is not known, and thus the integration

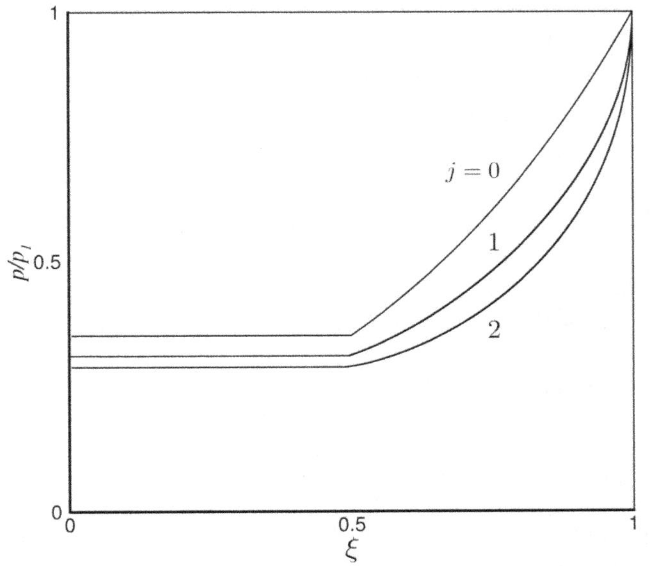

Figure 5.3 Pressure Profiles for Planar, Cylindrical, and Spherical C–J Detonation

cannot proceed from the shock $\xi = \frac{r}{R_s} = 1$. However, if we assume a shock strength M_s (or η), $\phi(1)$, and $\beta(1)$ can be determined from the Rankine–Hugoniot equations for $\phi(1)$ and $\beta(1)$, that is, Eqs 5.1.8 and 5.1.9. Equations 5.1.6 and 5.1.7 can then be integrated from $\xi = 1$ to $\xi = \xi_p$ (where ξ_p corresponds to the piston surface and $(\phi_p - \xi_p) = 0$).

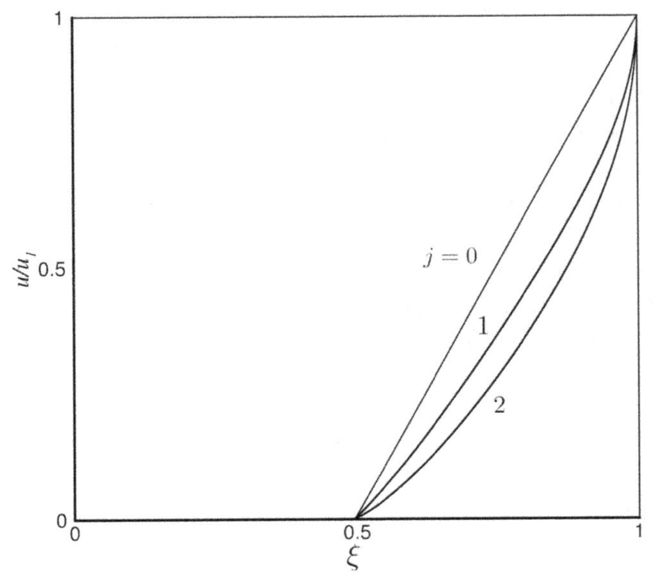

Figure 5.4 Particle Velocity Profiles for Planar, Cylindrical, and Spherical C-J Detonation

For strong shocks and values of $\gamma \simeq 1$, the distance between the shock and the piston is very small. An accurate approximate solution can be obtained by writing $\phi(\xi)$ and $\beta(\xi)$ as a power series in $(1-\xi)$, that is,

$$\phi(\xi) = \phi^{(0)} + \phi^{(1)}(1-\xi) + \cdots$$

$$\beta(\xi) = \beta^{(0)} + \beta^{(1)}(1-\xi) + \cdots$$

It is clear that $\phi^{(0)}$ and $\beta^{(0)}$ are just the boundary condition at the shock given by Eqs 5.1.8 and 5.1.9.

Substituting the expansion into the similarity equations (Eqs 5.1.6 and 5.1.7), the first-order coefficients $\phi^{(1)}$ and $\beta^{(1)}$ can readily be determined. Thus

$$\phi(\xi) = \phi(1) - \frac{j\phi(1)\beta^2(1)}{(\phi(1)-\xi)^2 - \beta^2(1)}(1-\xi) + \cdots \tag{5.4.1}$$

$$\beta(\xi) = \beta(1) - \frac{\gamma-1}{2}\frac{j\phi(1)\beta(1)\,(\phi(1)-1)}{(\phi(1)-1)^2 - \xi^2}(1-\xi) + \cdots, \tag{5.4.2}$$

where $\phi(1)$ and $\beta(1)$ are given by Eqs 5.1.8 and 5.1.9. Numerical integration of Eqs 5.1.6 and 5.1.7 for a constant velocity piston driven shock for the spherical case $j = 2$ are shown in Fig. 5.5. We note that the solution for $\phi(\xi)$ and $\beta(\xi)$ are very close to a linear form for strong shocks. Thus taking only the first order in the expansion $(1-\xi)$ (i.e., Eqs 5.4.1 and 5.4.2) can provide an accurate approximate solution.

Piston motion behind C-J detonation waves has been investigated by Lee et al. (1965) and Lee (1965, 1972). For diverging cylindrical or spherical Chapman–Jouguet detonations, we note from Eqs 5.3.9 and 5.3.10 that there are two possible solutions behind the C-J detonation front. For the compression solution where $\phi(\xi)$ and $\beta(\xi)$ increase with decreasing ξ behind the detonation, the solution terminates when $(\phi-\xi)=0$ which corresponds to a piston surface. For piston velocities greater than the critical value, the detonation becomes overdriven. For the expansion solution where ϕ and β decreases behind the detonation front, a constant velocity piston following the detonation will drive a shock wave in the detonation products. $\phi(\xi)$ and $\beta(\xi)$ will first decrease behind the Chapman–Jouguet detonation front, then undergo an abrupt increase across the shock. $\phi(\xi)$ and $\beta(\xi)$ will then continue to increase behind the shock until the piston surface where $(\phi-\xi)=0$. Figure 5.6 illustrates the wave trajectories and the conditions behind the detonation and the shock.

Across the CJ detonation, $\phi(1)$ and $\beta(1)$ are given by Eqs 5.3.1 and 5.3.2. Across the shock, the conservation equations in the frame of reference of the shock as shown below can be written as

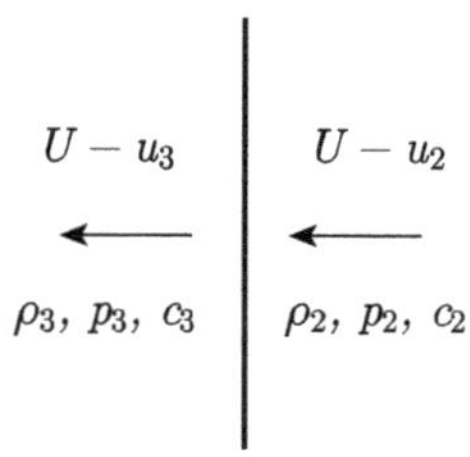

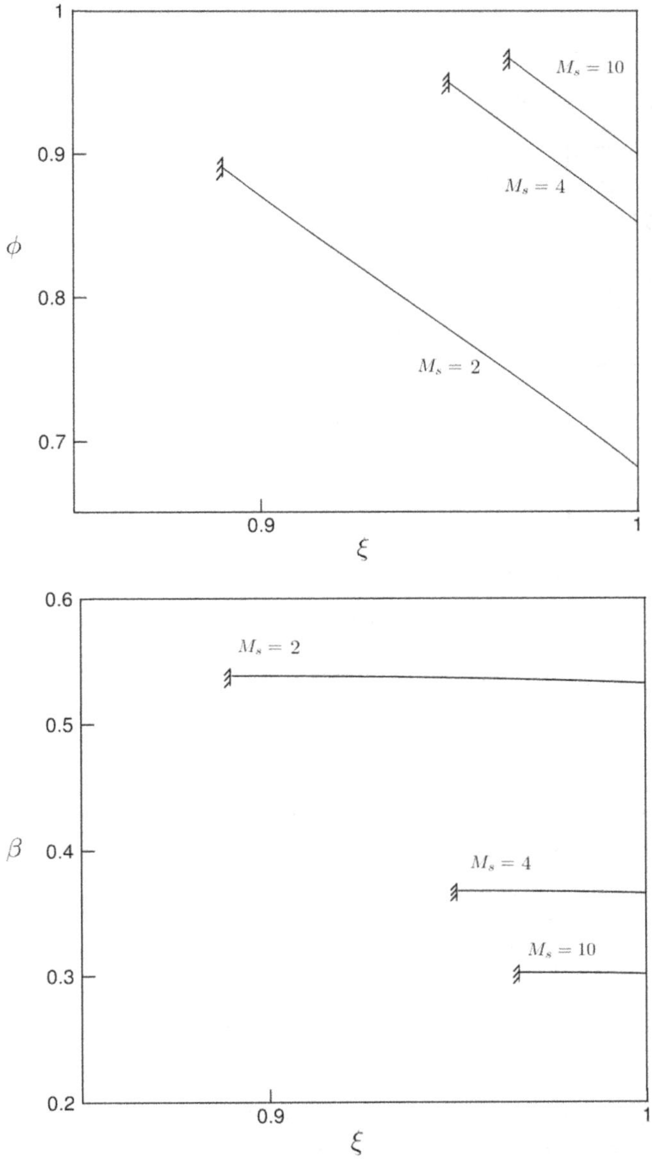

Figure 5.5 The Profiles of ϕ and β Behind a Shock Wave for Spherical Case

$$\rho_2(U - u_2) = \rho_3(U - u_3), \tag{5.4.3}$$

$$p_2 + \rho_2(U - u_2)^2 = p_3 + \rho_3(U - u_3)^2, \tag{5.4.4}$$

$$h_2 + \frac{(U - u_2)^2}{2} = h_3 + \frac{(U - u_3)^2}{2}. \tag{5.4.5}$$

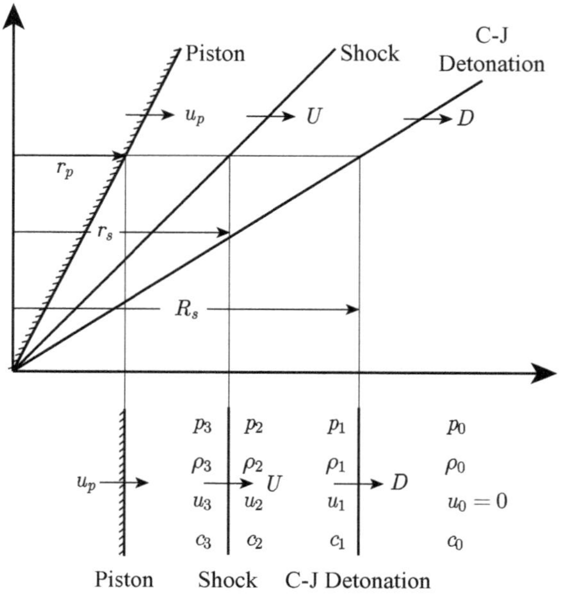

Figure 5.6 Conditions Behind Detonation and Shock in a Piston Driven Detonation

From Eq 5.4.3 the density ratio across the shock is

$$\frac{\rho_2}{\rho_3} = \frac{U - u_3}{U - u_2}. \tag{5.4.6}$$

Since the density ratio across the shock can be expressed in terms of the shock Mach number (or $\eta = \frac{1}{M_s^2}$), we write

$$\frac{\rho_2}{\rho_3} = \frac{(\gamma - 1) + 2\eta_s}{\gamma + 1}, \tag{5.4.7}$$

where

$$\eta_s = \frac{c_2^2}{(U - u_2)^2} = \frac{\beta_2^2}{(\xi_s - \phi_2)^2}, \tag{5.4.8}$$

and $\beta_2 = \frac{c_2}{D}$, $\phi_2 = \frac{u_2}{D}$, $\xi_s = \frac{r_s}{R_s} = \frac{\dot{r}_s t}{\dot{R}_s t} = \frac{Ut}{Dt} = \frac{U}{D}$. Note that r_s is the shock position whereas R_s is the position of the CJ detonation front. Equating Eqs 5.4.6 and 5.4.7 gives

$$\frac{\rho_2}{\rho_3} = \frac{\xi_s - \phi_3}{\xi_s - \phi_2} = \frac{(\gamma - 1) + 2\eta_s}{\gamma + 1}. \tag{5.4.9}$$

To obtain a solution, we use the boundary conditions (i.e., Eqs 5.3.1 and 5.3.2) for $\phi(1)$ and $\beta(1)$ at the CJ detonation front $\xi = 1$. Using Eqs 5.3.9 and 5.3.10 to compute the values of ϕ and β for a value of ξ close to the front, (e.g., $\xi = 0.999$), we can start the numerical integration of Eqs 5.1.6 and 5.1.7 for decreasing value of ξ. Note that the negative sign is used in Eqs 5.3.9 and 5.3.10 for an expansion solution. The integration

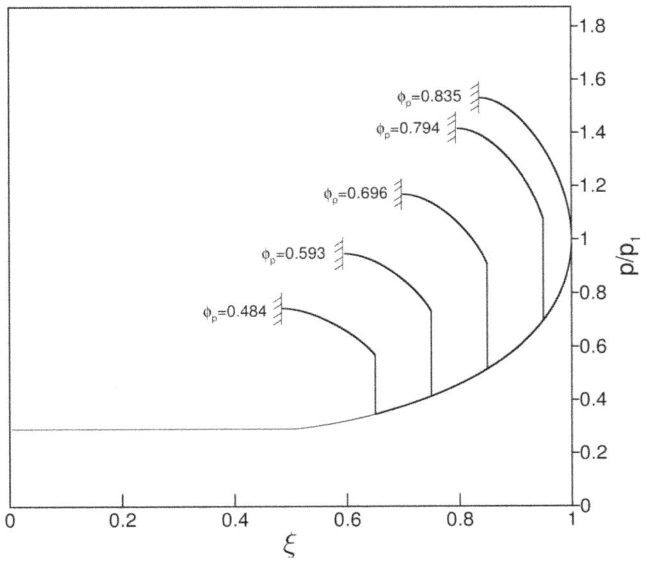

Figure 5.7 Pressure Distribution Behind a Spherical C-J Detonation for Various Piston Velocities

will proceed up to a desired value of ξ_s (hence the strength of the shock is known since $\xi_s = \frac{U}{D}$). The values for ϕ_2 and β_2 are obtained from the integration when $\xi = \xi_s$. η_s can be determined from Eq 5.4.9 since ξ_s, ϕ_2 are known and ϕ_s is given in terms of η_s from the Rankine–Hugoniot equations, that is,

$$\phi_3 = \frac{2}{\gamma+1}(1-\eta_s),$$

β_3 can be obtained from Eq 5.1.9, that is,

$$\beta_3 = \left(\frac{1}{\gamma+1}\right)[(2\gamma-(\gamma-1)\eta_s)(\gamma-1+2\eta_s)]^{\frac{1}{2}}.$$

With ϕ_3 and β_3 behind the shock determined, integration of Eqs 5.1.6 and 5.1.7 can be continued from ξ_s until $(\phi_p - \xi_p) = 0$ when the piston face is reached.

Figure 5.7 shows the pressure distribution behind a spherical detonation for various piston velocities. Note that, there corresponds a maximum piston velocity where ϕ and β decrease continuously to the CJ detonation front. In this case the secondary shock wave strength vanishes as it merges with the CJ detonation front. As the piston velocity decreases from this maximum value, the strength of the secondary shock increases to a maximum and then decreases again for further decrease in the piston velocity. When the piston velocity is zero, the shock vanishes and the flow expands continuously until $\phi \rightarrow 0$, $\beta \rightarrow$ constant, corresponding to the freely expanding Chapman–Jouguet cylindrical or spherical detonation.

Figure 5.8 shows the variation of the strength of the secondary shock with the piston velocity for a spherical C-J detonation. We note the strength of the shock rises rapidly

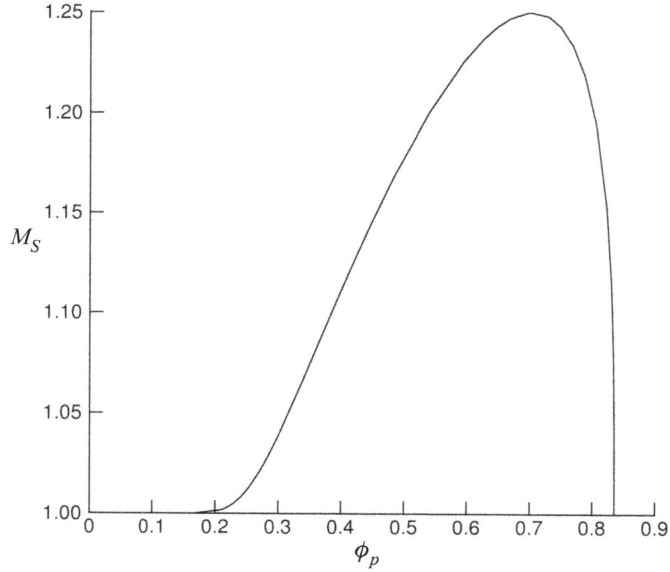

Figure 5.8 Variation of the Secondary Shock Velocity with Piston Velocity for a Spherical Detonation

with decreasing piston velocity to the maximum value. The rapid rise is due to the rapid decrease in the velocity behind the detonation.

For the planar geometry, the solution for $\phi(\xi)$ and $\beta(\xi)$ are given by Eqs 5.3.3 and 5.3.4. The strength of the shock M_s is given by

$$M_s = \frac{U - u_2}{c_2} = \frac{\xi_s - \phi_2}{\beta_2}. \tag{5.4.10}$$

Substituting the solution for ϕ_2 and β_2 from Eqs 5.3.3 and 5.3.4 into the above gives $M_s = 1$. Thus, for a planar piston following behind a planar Chapman–Jouguet detonation, $\phi(\xi)$ and $\beta(\xi)$ decrease linearly until $\xi = \xi_s$ and remain constant thereafter till the piston surface $(\phi - \xi) = 0$. Figure 5.9 shows the pressure distribution behind a planar CJ detonation for various piston velocities. Note that there is no secondary shock in contrast to the cylindrical and spherical cases.

For cylindrical and spherical geometries, we may also investigate the strength of the shock in the vicinity of the detonation front using the perturbation solution given by Eqs 5.3.9 and 5.3.10, that is,

$$\phi(\xi) = \phi(1) - \sqrt{\frac{2j\phi(1)\beta(1)}{\gamma + 1}}(1 - \xi)^{\frac{1}{2}},$$

$$\beta(\xi) = \beta(1) - \frac{\gamma - 1}{2}\sqrt{\frac{2j\phi(1)\beta(1)}{\gamma + 1}}(1 - \xi)^{\frac{1}{2}}.$$

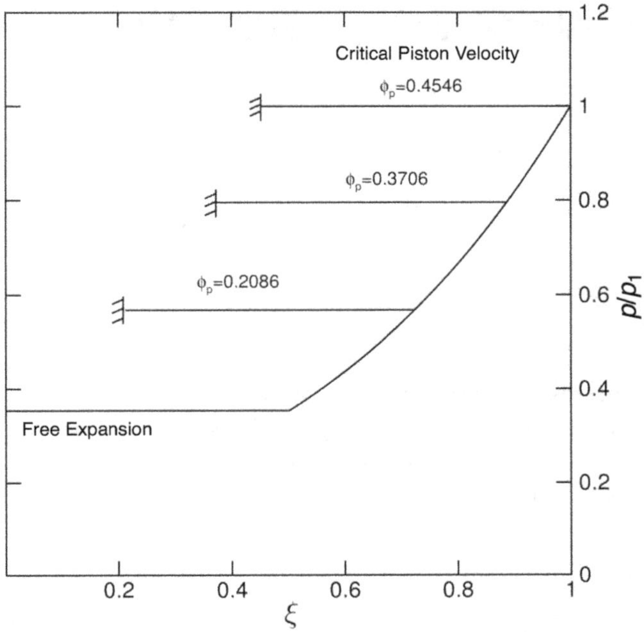

Figure 5.9 Pressure Distribution Behind a Planar Detonation for Various Piston Velocities

Substituting the above equation into Eq 5.4.10 gives

$$M_s = \frac{\xi_s - \phi_2}{\beta_2} = \frac{\xi_s - \phi(1) + \sqrt{\frac{2j\phi(1)\beta(1)}{\gamma+1}}(1-\xi_s)^{\frac{1}{2}}}{1 - \phi(1) + \sqrt{\frac{2j\phi(1)\beta(1)}{\gamma+1}}(1-\xi_x)^{\frac{1}{2}}},$$

where we have used the Chapman–Jouguet condition $\phi(1) + \beta(1) = 1$. Thus when $\xi_s \to 1$, $M_s \to 1$ and M_s increases as ξ_s decreases. Near $\xi = \xi_0$ when $\phi \to 0$ and $\beta \to$ constant, the solution may be written as

$$\phi(\xi) = \phi^{(1)}(\xi - \xi_0) + \cdots$$

$$\beta(\xi) = \beta^{(0)} + \beta^{(1)}(\xi - \xi_0) + \cdots$$

and in the vicinity of $\xi = \xi_0$, the shock strength can also be obtained as

$$M_s = \frac{\xi_s - \phi^{(1)}(\xi_s - \xi_0)}{\xi_0 - \beta^{(1)}(\xi_s - \xi_0)}.$$

Thus when $\xi \to \xi_0$, $M_s \to 1$ and the shock degenerates into a Mach wave. Away from $\xi = 1$ and $\xi = \xi_0$, ϕ_2 and β_2 must be determined from the numerical solution of Eqs 5.1.6 and 5.1.7.

6 The Snow-Plow Approximation

6.1 Introduction

The density ratio across a strong shock wave is given by $\frac{\rho_1}{\rho_0} = \frac{\gamma+1}{\gamma-1}$. Thus for values of $\gamma \simeq 1$, $\frac{\rho_1}{\rho_0} \gg 1$ and all the mass swept up by the shock wave will be concentrated in a thin layer at the shock front. To estimate of the thickness δ of this dense shock layer, consider a spherical strong shock at radius, R_s. Mass balance gives

$$\frac{4}{3}\pi R_s^3 \rho_0 = 4\pi R_s^2 \delta \rho_1,$$

and thus $\frac{\delta}{R_s} = \frac{1}{3}\frac{\rho_0}{\rho_1} = \frac{1}{3}\frac{\gamma-1}{\gamma+1}$. For $\gamma = 1.2$, $\frac{\delta}{R_s} = 0.03$, that is, δ is only 3% of the shock radius. Even for $\gamma = 1.4$, $\frac{\delta}{R_s} = 0.055$ and δ is only 5% of the shock radius. Thus the shock acts like a "snow plow" sweeping up the mass in its path and concentrating it in a thin layer at the shock front. The velocity of the mass behind the shock is given by $\frac{u_1}{\dot{R}_s} = \frac{2}{\gamma+1}$ and for $\gamma \simeq 1$, $u_1 \simeq \dot{R}_s$. Thus the acceleration of the material is close to the acceleration of the shock. Writing the pressure difference across the shock layer then gives the acceleration of the shock wave using Newton's law. With most of the mass concentrated at the shock front, very little mass remains in the core. With modest acceleration of the mass in the core, the pressure will be uniform in the low density core. The temperature however will be high and hence also the internal energy. However, the kinetic energy of the low density core is small. Based on the snow-plow approximation, an equation for the shock acceleration, the shock pressure, and the pressure of the core can be obtained. Together with the Rankine–Hugoniot relationship between the shock pressure and the shock velocity, and the conservation of total energy (energy integral), the shock trajectory can be found. The snow-plow approximation was used by Bethe (1947) to describe the propagation of strong blast waves. Later it was used by Rosenbluth and Garwin (1954) for shock waves driven by current sheets. Chernyi (1961), Cole (1957), and Freeman (1962) also used a similar concept (the Newtonian approximation) for the description of hypersonic blunt body flows. The analysis presented in this chapter follows closely the work of Bethe (1947).

6.2 Basic Equations

Following Chernyi (1961), we shall describe a more systematic approach of successive approximation based on a power series expansion of the solution in a small parameter

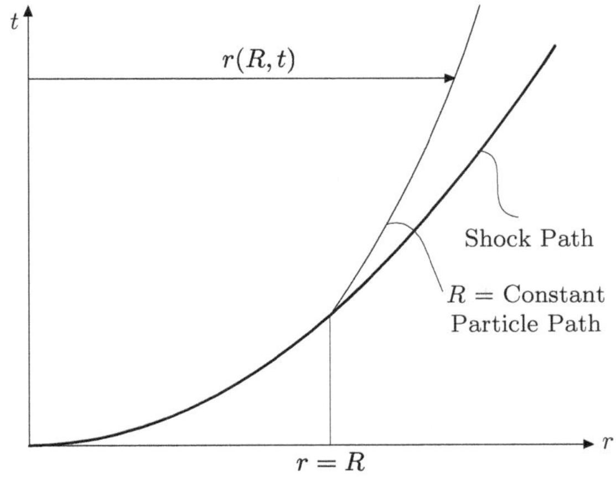

Figure 6.1 Fluid Particle Having Initial Position R Entering the Shock

$\varepsilon = \frac{\gamma-1}{\gamma+1}$. The Lagrangian form of the conservation equations for planar ($j = 0$), cylindrical ($j = 1$) and spherical ($j = 2$) symmetries as given in Chapter 1 are as

$$\left.\frac{\partial r}{\partial R}\right)_t = \frac{\rho_0}{\rho}\left(\frac{R}{r}\right)^j, \tag{6.2.1}$$

$$\left.\frac{\partial u}{\partial t}\right)_R = \left.\frac{\partial^2 r}{\partial t^2}\right)_R = -\frac{1}{\rho_0}\left(\frac{r}{R}\right)^j\left(\frac{\partial p}{\partial R}\right)_t, \tag{6.2.2}$$

$$\left.\frac{\partial s}{\partial t}\right)_R = \frac{\partial}{\partial t}\left(\frac{p}{\rho^\gamma}\right)_R = 0. \tag{6.2.3}$$

The independent variables are the time t, and the Lagrangian coordinate R of a fluid particle. This is shown in Fig. 6.1. We choose $r = R$ to be the initial position of a fluid particle. We seek solutions for the dependent variables $r(R, t)$, $\rho(R, t)$, $p(R, t)$ in the form of a power series in $\varepsilon = \frac{\gamma-1}{\gamma+1}$, that is,

$$\left.\begin{aligned} r(R,t) &= R_s(t) + \varepsilon R^{(1)} + \varepsilon^2 R^{(2)} + \cdots \\ \rho(R,t) &= \frac{1}{\varepsilon}\rho^{(0)} + \rho^{(1)} + \varepsilon\rho^{(2)} + \cdots \\ p(R,t) &= p^{(0)} + \varepsilon p^{(1)} + \varepsilon^2 p^{(2)} + \cdots \end{aligned}\right\} \tag{6.2.4}$$

Substituting the expansions (Eq 6.2.4) into the basic equations and collecting terms of the same order in ε yield

Zeroth order in ε

$$\left.\frac{\partial R_s(t)}{\partial R}\right)_t = 0, \tag{6.2.5}$$

$$\left.\frac{\partial^2 R_s}{\partial t^2}\right)_R = -\frac{1}{\rho_0}\left(\frac{R_s}{R}\right)^j \left(\frac{\partial p^{(0)}}{\partial R}\right)_t, \tag{6.2.6}$$

$$\frac{\partial}{\partial t}\left(\frac{p^{(0)}}{\rho^{(0)\gamma}}\right)_R = 0. \tag{6.2.7}$$

First order in ε

$$\left.\frac{\partial R^{(1)}}{\partial R}\right)_t = \left(\frac{R}{R_s}\right)^j \left(\frac{\rho_0}{\rho^{(0)}}\right), \tag{6.2.8}$$

$$\left.\frac{\partial^2 R^{(1)}}{\partial t^2}\right)_R = -\frac{1}{\rho_0}\left(\frac{R_s}{R}\right)^j \left(j\left(\frac{R^{(1)}}{R_s}\right)\frac{\partial p^{(0)}}{\partial R} + \frac{\partial p^{(1)}}{\partial R}\right)_t, \tag{6.2.9}$$

$$\frac{\partial}{\partial t}\left(\frac{p^{(1)}}{p^{(0)}} - \frac{\gamma\rho^{(1)}}{\rho^{(0)}}\right)_R = 0, \tag{6.2.10}$$

and similarly we can continue to obtain the equations for higher order in ε.

The boundary conditions at the strong shock front are given as

$$\frac{\rho_1}{\rho_0} = \frac{\gamma+1}{\gamma-1}, \qquad \frac{u_1}{\dot{R}_s} = \frac{p_1}{\rho_0\dot{R}_s^2} = \frac{2}{\gamma+1}. \tag{6.2.11}$$

With $\varepsilon = \frac{\gamma-1}{\gamma+1}$, we write

$$\rho_1 = \frac{\rho_0}{\varepsilon} = \frac{\rho_1^{(0)}}{\varepsilon} + \rho_1^{(1)} + \varepsilon\rho_1^{(2)} + \cdots$$

and thus

$$\begin{aligned}\rho_1^{(0)} &= \rho_0,\\ \rho_1^{(1)} &= 0.\end{aligned} \tag{6.2.12}$$

For the pressure, we write

$$p_1 = \frac{2}{\gamma+1}\rho_0\dot{R}_s^2 = \rho_0\dot{R}_s^2(1-\varepsilon) = p_1^{(0)} + p_1^{(1)} + \cdots$$

Thus

$$\begin{aligned}p_1^{(0)} &= \rho_0\dot{R}_s^2,\\ p_1^{(1)} &= -\rho_0\dot{R}_s^2,\end{aligned} \tag{6.2.13}$$

and similarly for the particle velocity

$$u_1 = u_1^{(0)} + \varepsilon u_1^{(1)} + \cdots = \frac{2}{\gamma+1}\dot{R}_s = \dot{R}_s(1-\epsilon),$$

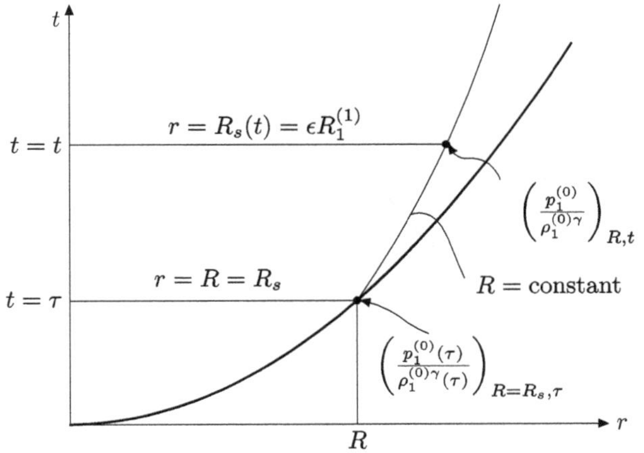

Figure 6.2 Expansion of Fluid Particle Entering with Lagrangian Coordinate R

and

$$\begin{aligned} u_1^{(0)} &= \dot{R}_s, \\ u_1^{(1)} &= -\dot{R}_s. \end{aligned} \tag{6.2.14}$$

We can now proceed with the integration of the zeroth-order equations. From Eq 6.2.5, we get R_s as a function of t and Eq 6.2.6 becomes

$$\left.\frac{\partial p^{(0)}}{\partial R}\right) = -\rho_0 \ddot{R}_s \left(\frac{R}{R_s}\right)^j,$$

which integrates to yield

$$p^{(0)} = -\frac{\rho_0 \ddot{R}_s R_s}{(j+1)} \left(\frac{R}{R_s}\right)^{j+1} + C(t).$$

The integration constant, $C(t)$, can be evaluated from the condition at the shock front $R = R_s$. Doing so, we obtain

$$p^{(0)} = \rho_0 \dot{R}_s^2 + \frac{\rho_0 \ddot{R}_s R_s}{(j+1)} \left(1 - \left(\frac{R}{R_s}\right)^{(j+1)}\right). \tag{6.2.15}$$

The solution for the zeroth-order density $\rho^{(0)}$ can be obtained from Eq 6.2.7 which integrates to yield

$$\left(\frac{p^{(0)}}{\rho^{(0)\gamma}}\right)_R = \text{constant}.$$

Evaluating the constant at the shock front gives

$$\left.\frac{p^{(0)}}{\rho^{(0)\gamma}}\right)_{R,t} = \left(\frac{p_1^{(0)}}{\rho_1^{(0)\gamma}}\right)_{R=R_s,\tau} = \frac{p_1^{(0)}(t=\tau)}{\rho_1^{(0)\gamma}(t=\tau)}.$$

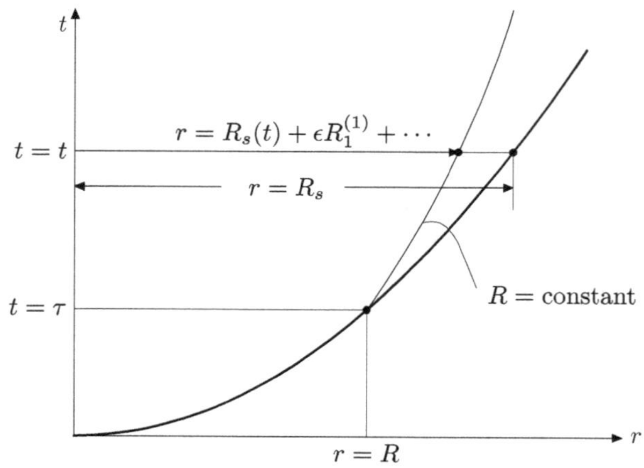

Figure 6.3 Fluid Particle Having Lagrangian Coordinate R Closely Following the Shock Trajectory

The above equation simply states that the entropy of a particle R at time t, that is, $\left(\frac{p^{(0)}}{\rho^{(0)\gamma}}\right)_{R,t}$ is equal to the value when it first crosses the shock $R = R_s$, at time $t = \tau$. Using the boundary conditions for $p_1^{(0)} = \rho_0 \dot{R}_s^2$, $\rho_1^{(0)} = \rho_0$, we get

$$\rho^{(0)}(R,t) = \rho_0 \left(\frac{\dot{R}_s^2(t)}{\dot{R}_s^2(\tau)}\right)^{\frac{1}{\gamma}} \left[1 + \frac{R_s \ddot{R}_s}{\dot{R}_s^2(j+1)} - \left(1 - \left(\frac{R}{R_s}\right)^{j+1}\right)\right]^{\frac{1}{\gamma}}. \tag{6.2.16}$$

Note that $\dot{R}_s(t)$ and $\dot{R}_s(\tau)$ represent the shock velocity at time t and at time τ. The zeroth-order solution for $r(R,t) = R_s(t)$ simply means that all the mass swept up by the shock is concentrated at the shock where $\varepsilon = 0$.

The zeroth-order solution is now complete and we proceed to the first order in ε. Integrating Eq 6.2.8, we write

$$\int_{R^{(1)}}^{0} dR^{(1)} = \int_{R}^{R_s} \frac{\rho_0}{\rho^{(0)}} \left(\frac{R}{R_s}\right)^j dR.$$

The integral is carried out at constant time t from the particle path $r = R = R_s + \varepsilon R^{(1)} + \cdots$ to the shock front $r = R_s$ as shown in Fig. 6.3.

Substituting the solution for $\rho^{(0)}$ into the above equation gives

$$R^{(1)} = -\int_{R}^{R_s} \left[\frac{\dot{R}_1^2(\tau)}{\dot{R}_s^2(t)}\right]^{\frac{1}{\gamma}} \frac{\left(\frac{R}{R_s}\right)^j \mathrm{d}R}{\left[1 + \frac{R_s \ddot{R}_s}{\dot{R}_s^2(j+1)}\left(1 - \left(\frac{R}{R_s}\right)^{j+1}\right)\right]^{\frac{1}{\gamma}}}, \tag{6.2.17}$$

where $t = \tau$ is when the particle first crosses the shock front $r = R = R_s(\tau)$.

The first-order solution for the pressure can be obtained by integrating Eq 6.2.9, that is,

$$p^{(1)} = -\rho_0 \dot{R}_s^2 + \int_R^{R_s} \left[\rho_0 \frac{\partial^2 R^{(1)}}{\partial t^2} \left(\frac{R}{R_s} \right)^j - \frac{j \rho_0 \ddot{R}_s R^j R^{(1)}}{R_s^{j+1}} \right] dR, \tag{6.2.18}$$

where we have used the boundary condition $p_0^{(1)} = -\rho_0 \dot{R}_s^2$. The first-order solution for the density $\rho^{(1)}$ can be obtained by integrating Eq 6.2.10, that is,

$$\frac{p^{(1)}}{p^{(0)}} - \frac{\gamma \rho^{(1)}}{\rho^{(0)}} = \text{constant}.$$

The constant can be evaluated at time τ where the particle R first crosses the shock. Thus

$$\frac{p^{(1)}(R,t)}{p^{(0)}(R,t)} - \frac{\gamma \rho^{(1)}(R,t)}{\rho^{(0)}(R,t)} = \frac{p_1^{(1)}(R_s,t)}{p_1^{(0)}(R_s,t)} - \frac{\gamma \rho_1^{(1)}(R_s,t)}{\rho_1^{(0)}(R_s,t)}.$$

Since $p_1^{(1)} = -\rho_0 \dot{R}_s^2$ and $\rho_1^{(1)} = 0$, the above equation yields

$$\rho^{(1)}(R,t) = \frac{\rho^{(0)}(R,t)}{\gamma} \left[\frac{p^{(1)}(R,t)}{p^{(0)}(R,t)} + 1 \right]. \tag{6.2.19}$$

The solutions for $p^{(1)}(R,t)$, $\rho^{(0)}(R,t)$, $p^{(0)}(R,t)$ are given by Eqs 6.2.18, 6.2.16, and 6.2.15, respectively.

The remaining task is to determine $R_s(t)$ and this requires an additional equation beside the conservation equations. The additional equation comes from the nature of the particular explosion problem itself. For example, for the constant energy blast problem, the conservation of total energy, that is, the energy integral, is used. For piston driven explosions, the piston path $R_p(t)$ has to be specified.

6.3 Constant Energy Blast Waves

The conservation of total energy at any instant when the blast wave is at a radius $R_s(t)$ can be written as

$$E_0 = \int_0^{R_s} \frac{\rho u^2}{2} k_j r^j dr + \int_0^{R_s} \frac{p}{\gamma - 1} k_j r^j dr,$$

when $k_j = 1, 2\pi, 4\pi$ for $j = 0, 1,$ and 2, respectively. We have neglected the initial internal energy of the medium when compared to internal energy of the shocked medium. $\frac{p}{\gamma-1}$ is the internal energy per unit volume for a perfect gas with constant γ. For the zeroth-order solution where all the mass is concentrated at the shock, $u \simeq u_1 \simeq \dot{R}_s$ and the kinetic energy term can be written as

$$\int_0^{R_s} \frac{\rho u^2}{2} k_j r^j dr = \frac{\rho_0 \dot{R}_s^2 R_s^{j+1}}{2(j+1)} k_j.$$

The internal energy term can be written as

$$\int_0^{R_s} \frac{p}{\gamma-1} k_j r^j dr = \int_0^{R_s-\delta} \frac{p_c}{\gamma-1} k_j r^j dr + \int_{R_s-\delta}^{R_s} \frac{p_1}{\gamma-1} k_j r^j dr,$$

where p is the pressure at the shock front and p_c is the pressure at the core of the blast wave.

For the lowest order, $p_1 \simeq \rho_0 \dot{R}_s^2$ and if we assume the pressure of the high temperature core $p_c \simeq \alpha p_1$, where α is some constant of the order of unity, the internal energy integral can then be written as

$$\int_0^{R_s} \frac{p}{\gamma-1} k_j r^j dr = \frac{\rho_0 \dot{R}_s^2 R_s^{j+1} k_j \alpha}{(\gamma-1)(j+1)} \left(1-\frac{\delta}{R_s}\right)^{j+1}$$

$$+ \frac{\rho_0 \dot{R}_s^2 R_s^{j+1}}{(\gamma-1)(j+1)} \left(1-\left(1-\frac{\delta}{R_s}\right)^{j+1}\right),$$

and for the lowest order of approximation where $\frac{\delta}{R} \simeq 0$, the above integral reduces to

$$\int_0^{R_s} \frac{p}{\gamma-1} k_j r^j dr \simeq \frac{\rho_0 \dot{R}_s^2 R_s^{j+1} k_j \alpha}{(\gamma-1)(j+1)}.$$

The energy integral can now be written as

$$E_0 = \frac{\rho_0 \dot{R}_s^2 R_s^{j+1} k_j}{2(j+1)} + \frac{\rho_0 \dot{R}_s^2 R_s^{j+1} \alpha k_j}{(\gamma-1)(j+1)}$$

$$= \frac{\rho_0 \dot{R}_s^2 R_s^{j+1}}{(j+1)} \left[\frac{1}{2} + \frac{\alpha}{\gamma-1}\right].$$

Since E_0 is constant, the right hand side cannot be time dependent explicitly. Thus $\dot{R}_s^2 R_s^{j+1} \to$ constant or $\dot{R}_s^2 \to \frac{1}{R_s^{j+1}}$ giving $\dot{R}_s = A R_s^{-\frac{j+1}{2}}$. The term $\frac{R_s \ddot{R}_s}{\dot{R}_s^2}$ then becomes

$$\frac{R_s \ddot{R}_s}{\dot{R}_s^2} = -\left(\frac{j+1}{2}\right). \tag{6.3.1}$$

The solution for the pressure at the core where $R \to 0$ from Eq 6.2.15 is given by

$$p(0,t) \simeq \rho_0 \dot{R}_s^2 \left(1 + \frac{R_s \ddot{R}_s}{\dot{R}_s^2 (j+1)}\right) = \frac{\rho_0 \dot{R}_s^2}{2}.$$

Thus $\alpha = \frac{1}{2}$ and the energy integral becomes

$$E_0 = \frac{\rho_0 k_j \dot{R}_s^2 R_s^{j+1} \gamma A^2}{2(j+1)(\gamma-1)}.$$

Solving for A gives

$$A = \left(\frac{2(j+1)(\gamma-1)E_0}{\rho_0 k_j \gamma}\right)^{\frac{1}{2}}. \tag{6.3.2}$$

The shock trajectory can be obtained by integrating $\dot{R}_s = AR_s^{-\frac{j+1}{2}}$, that is,

$$\dot{R}_s = \frac{dR_s}{dt} = AR_s^{-\frac{j+1}{2}},$$

hence

$$R_s = \left[\frac{j+3}{2}A\right]^{\frac{2}{j+3}} t^{\frac{2}{j+3}}, \tag{6.3.3}$$

and substituting the expression for A into the above gives

$$R_s(t) = \left[\left(\frac{j+3}{2}\right)^2 \frac{2(j+1)(\gamma-1)}{\gamma k_j}\frac{E_0}{\rho_0}\right]^{\frac{1}{j+3}} t^{\frac{2}{j+3}},$$

or in the following form

$$R_s(t) = B\left(\frac{E_0}{\rho_0}\right)^{\frac{1}{j+3}} t^{\frac{2}{j+3}}. \tag{6.3.4}$$

For spherical blast ($j = 2$), the constant B for $\gamma = 1.4$ and 1.2 is found to be 0.9686 and 0.869, respectively. The corresponding values from the exact self-similar solution are 1.033 and 0.89. Thus, we note the accuracy of the snow-plow solution is very good even considering just the lowest order.

Using Eq 6.3.1, the zeroth-order pressure distribution can be written as

$$p^{(0)}(R,t) = \frac{\rho_0 \dot{R}_s^2}{2}\left(1 + \left(\frac{R}{R_s}\right)^{j+1}\right), \tag{6.3.5}$$

and at the shock front where $R = R_s$, we see that $p^{(0)} \to p_1^{(0)} = \rho_0 \dot{R}_s^2$ in accord with the boundary conditions given by Eq 6.2.13. The zeroth-order solution for the density given by Eq 6.2.16 becomes

$$\rho^{(0)}(R,t) = \rho_0\left[\frac{1}{2}\left(\frac{\dot{R}_s^2(t)}{\dot{R}_s^2(\tau)}\right)\left(1 + \left(\frac{R}{R_s}\right)^{j+1}\right)\right]^{\frac{1}{\gamma}},$$

where $\dot{R}_s(\tau)$ corresponds to the shock velocity when the particle first crosses the shock front, that is, $R = R_s$. Since $\dot{R}_s^2 \sim \frac{1}{R_s^{j+1}}$ we can write $\frac{\dot{R}_s^2(t)}{\dot{R}_s^2(\tau)}$ as $\left(\frac{R}{R_s}\right)^{j+1}$, thus

$$\rho^{(0)}(R,t) = \frac{\rho_0}{2}\left[\left(\frac{R}{R_s}\right)^{j+1}\left(1 + \left(\frac{R}{R_s}\right)^{j+1}\right)\right]^{\frac{1}{\gamma}}. \tag{6.3.6}$$

It is of interest to obtain the pressure and density distributions in terms of the Eulerian coordinates $\frac{r}{R_s}$. From Eq 6.2.1, that is,

$$\left.\frac{\partial r}{\partial R}\right)_t = \frac{\rho_0}{\rho}\left(\frac{R}{r}\right)^j = \frac{\rho_0}{\frac{\rho^{(0)}}{\varepsilon}}\left(\frac{R}{r}\right)^j,$$

we get

$$\left(\frac{r}{R}\right)^j\left(\frac{\partial r}{\partial R}\right)=\frac{\varepsilon\rho_0}{\rho^{(0)}}=\varepsilon\left[\frac{2}{\left(\frac{R}{R_s}\right)^{j+1}\left(1+\left(\frac{R}{R_s}\right)^{j+1}\right)}\right]^{\frac{1}{\gamma}}.$$

Defining $x=\left(\frac{R}{R_s}\right)^{j+1}$ and $y=\left(\frac{r}{R_s}\right)^{j+1}$, the above equation becomes

$$\frac{\mathrm{d}y}{\mathrm{d}x}=\left(\frac{r}{R}\right)^j\left(\frac{\partial r}{\partial R}\right)_t=\frac{\rho_0}{\rho}=\varepsilon\left(\frac{2}{x(1+x)}\right)^{\frac{1}{\gamma}}. \tag{6.3.7}$$

Since $\varepsilon=\frac{\gamma-1}{\gamma+1}$, and $\delta=\gamma-1$, we have $\varepsilon=\frac{\delta}{2+\delta}=\frac{\delta}{2}\left(1-\frac{\delta}{2}+\cdots\right)$ and Eq 6.3.7 becomes

$$\frac{\mathrm{d}y}{\mathrm{d}x}=\frac{\gamma-1}{\gamma+1}\left(\frac{2}{x(1+x)}\right)^{\frac{1}{\gamma}}=\frac{\delta}{2+\delta}\left(\frac{2}{x(1+x)}\right)^{\frac{1}{1+\delta}}.$$

Since $\delta\ll1$, we may expand the above equation and write

$$\frac{\mathrm{d}y}{\mathrm{d}x}=\frac{\delta}{2}\left(1-\frac{\delta}{2}+\cdots\right)\left(\frac{2}{x(1+x)}\right)^{1-\delta+\cdots}. \tag{6.3.8}$$

We can expand the last term on the right hand side of the above as follows: Let $\alpha=\frac{2}{x(1+x)}$, thus $\left(\frac{2}{x(1+x)}\right)^{1-\delta+\cdots}=\alpha^{1-\delta+\cdots}=\alpha\alpha^{-\delta}=\alpha e^{\ln\alpha^{-\delta}}$ and the exponential term can be expanded as

$$e^{\ln\alpha^{-\delta}}=1+\ln\alpha^{-\delta}+O\left[\delta^2\right]=1-\delta\ln x.$$

Thus Eq 6.3.8 becomes

$$\frac{\mathrm{d}y}{\mathrm{d}x}=\frac{\delta}{2}\left(1-\frac{\delta}{2}+\cdots\right)\left(\frac{2}{x(1+x)}\right)\left(1-\delta\ln\frac{2}{x(1+x)}+\cdots\right),$$

and retaining first-order terms only, we get

$$\frac{\mathrm{d}y}{\mathrm{d}x}=\frac{\delta}{x(1+x)}. \tag{6.3.9}$$

In neglecting terms of the order of δ^2, we imply

$$\frac{2}{x(1+x)}\gg-\delta\frac{2}{x(1+x)}\ln\frac{2}{x(1+x)}$$

or

$$e^{\frac{1}{\delta}}\gg\frac{x(1+x)}{2}.$$

Since $\delta\ll1$, the above condition is usually satisfied with $x\sim O[1]$. Integrating Eq 6.3.9, that is,

$$\int_1^y\mathrm{d}y=\int_1^x\frac{\delta}{x}\mathrm{d}x-\int_1^x\frac{\delta}{1+x}\mathrm{d}x,$$

we obtain,

$$y = 1 - (\gamma - 1)\ln\frac{(1+x)}{2x},$$

thus

$$\left(\frac{r}{R_s}\right)^{j+1} = 1 - (\gamma - 1)\ln\frac{R_s^{j+1} + R^{j+1}}{2R^{j+1}}. \tag{6.3.10}$$

For $x \ll 1$, we can write Eq 6.3.7 as

$$\frac{\mathrm{d}y}{\mathrm{d}x} = \varepsilon\left(\frac{2}{x(1+x)}\right)^{\frac{1}{\gamma}} \simeq \varepsilon\left(\frac{2}{x}\right)^{\frac{1}{\gamma}},$$

where we have neglected x^2 as compared to x. Integrating the above equation yields

$$y = \frac{2^{\frac{1}{\gamma}}}{\gamma+1}x^{\frac{\gamma-1}{\gamma}} + \text{constant},$$

and since $\gamma \simeq 1$, the above equation can be written as

$$y \simeq x^{\frac{\gamma-1}{\gamma}} + \text{constant}. \tag{6.3.11}$$

Note that $x^{\frac{\gamma-1}{\gamma}}$ can be written as

$$x^{\frac{\gamma-1}{\gamma}} = e^{\frac{\gamma-1}{\gamma}\ln x} = 1 + \frac{\gamma-1}{\gamma}\ln x + \cdots$$

and comparing with Eq 6.3.10, that is,

$$\begin{aligned} y &= 1 - (\gamma - 1)\ln\frac{(1+x)}{2x}, \\ &= 1 - (\gamma - 1)\ln(1+x) + (\gamma - 1)\ln 2 + (\gamma - 1)\ln x, \end{aligned}$$

which for $x \ll 1$, $1 + x \simeq 1$ and neglecting terms of the order of $(\gamma - 1)$, we get

$$y = 1 + (\gamma - 1)\ln x,$$

which is the same as Eq 6.3.11, that is,

$$y \simeq x^{\frac{\gamma-1}{\gamma}} + \text{constant} = 1 + \frac{(\gamma-1)}{\gamma}\ln x - \cdots$$

with $\gamma \simeq 1$. Comparing Eqs 6.3.10 and 6.3.11, we note that the constant is equal to zero and so we write the solution for y as

$$y \simeq x^{\frac{\gamma-1}{\gamma}}$$

or

$$\left(\frac{r}{R_s}\right)^{j+1} \simeq \left(\frac{R}{R_s}\right)^{\left(\frac{\gamma-1}{\gamma}\right)(j+1)},$$

hence

$$\left(\frac{r}{R_s}\right) = \left(\frac{R}{R_s}\right)^{\frac{\gamma-1}{\gamma}}. \tag{6.3.12}$$

Since $(\gamma - 1) \ll 1$, any fluid particle defined by a value of R will be very close to the shock front, that is, $r(R, t) \simeq R_s(t)$.

For the particle velocity, we differentiate Eq 6.3.12 with respect to time, that is,

$$u = \left(\frac{\partial r}{\partial t}\right)_R = \frac{\dot{R}_s}{\gamma}\left(\frac{R}{R_s}\right)^{\frac{\gamma-1}{\gamma}},$$

and since $\gamma \simeq 1$, we obtain

$$u \simeq \dot{R}_s\left(\frac{r}{R}\right), \tag{6.3.13}$$

using Eq 6.3.12. Thus the velocity profile is linear.

The density profile can be obtained from Eq 6.2.1, that is,

$$\left(\frac{r}{R}\right)^j\left(\frac{\partial r}{\partial R}\right)_t = \frac{\rho_0}{\rho} = \frac{\varepsilon\rho_0}{\rho^{(0)}} = \varepsilon\left[\frac{2}{\left(\frac{R}{R_s}\right)^{j+1}\left(1+\left(\frac{R}{R_s}\right)^{j+1}\right)}\right]^{\frac{1}{\gamma}},$$

where we have used Eq 6.2.16 for the solution of $\rho^{(0)}$. Since $x = \left(\frac{R}{R_s}\right)^{j+1}$ and $y = \left(\frac{r}{R_s}\right)^{j+1}$, the above equation becomes

$$\left(\frac{r}{R}\right)^j\left(\frac{\partial r}{\partial R}\right) = \frac{\mathrm{d}u}{\mathrm{d}x} = \varepsilon\left[\frac{2}{x(1+x)}\right]^{\frac{1}{\gamma}} \simeq \varepsilon\left(\frac{2}{x}\right)^{\frac{1}{\gamma}}.$$

With $u = x^{\frac{\gamma-1}{\gamma}}$ (or $x = y^{\frac{\gamma}{\gamma-1}}$), the above equation becomes

$$\frac{\rho_0}{\rho} = \frac{\rho_0}{\rho_1}\left(\frac{2}{y^{\frac{1}{\gamma-1}}}\right) \simeq \frac{\rho_0}{\rho_1}\left(\frac{2}{y^{\frac{1}{\gamma-1}}}\right),$$

where we have replaced ε by $\frac{\rho_0}{\rho_1}$. Since $y = \left(\frac{r}{R_s}\right)^{j+1}$, the above equation becomes

$$\frac{\rho}{\rho_0} = \frac{1}{2}\frac{\rho_1}{\rho_0}\left(\frac{r}{R_s}\right)^{\frac{j+1}{\gamma-1}}, \tag{6.3.14}$$

which gives a power law profile for the density. Since $(\gamma - 1) \ll 1$, the exponent of $\frac{r}{R_s}$ is very large. Thus the density decreases very rapidly behind the shock, that is, all the mass is concentrated near the shock.

In the exact similarity solution for constant energy blast waves, it is found that the density profile behind the shock can be approximated by a simple power law of the form

$$\frac{\rho}{\rho_0} = \frac{\rho_1}{\rho_0}\left(\frac{r}{R_s}\right)^q.$$

The exponent q can be evaluated from the mass integral, that is,

$$\int_0^{R_s}\rho_0 k_j r^j \mathrm{d}r = \frac{\rho_0 k_j R_s^{j+1}}{j+1} = \int_0^{R_s}\rho_1\left(\frac{r}{R_s}\right)^q k_j r^j \mathrm{d}r = \frac{\rho_1 k_j R_s^{j+1}}{q+j+1},$$

which gives

$$q = (j+1)\left(\frac{\rho_1}{\rho_0} - 1\right) = \frac{2(j+1)}{\gamma - 1},$$

when $\frac{\rho_1}{\rho_0} = \frac{\gamma+1}{\gamma-1}$ is substituted into the above equation. The density profile is thus

$$\frac{\rho}{\rho_0} = \frac{\rho_1}{\rho_0}\left(\frac{r}{R_s}\right)^{\frac{2(j+1)}{\gamma-1}}, \tag{6.3.15}$$

which is very similar to Eq 6.3.14 of the zeroth-order snow-plow solution.

The pressure profile is given by Eq 6.3.5 as

$$p^{(0)} = \frac{\rho_0 \dot{R}_s^2}{2}\left(1 + \left(\frac{R}{R_s}\right)^{j+1}\right).$$

Using Eq 6.3.12, the above equation can be written in terms of the Eulerian coordinate $\left(\frac{r}{R_s}\right)$ as

$$p^{(0)} = \frac{\rho_0 \dot{R}_s^2}{2}\left(1 + \left(\frac{r}{R_s}\right)^{(j+1)\left(\frac{\gamma}{\gamma-1}\right)}\right), \tag{6.3.16}$$

which is also a power law decreasing rapidly from $\rho_0 \dot{R}_s^2$ at the shock front to $\frac{\rho_0 \dot{R}_s^2}{2}$ as $r \to 0$.

With the zeroth-order solution determined one can proceed to higher orders. However, the complexity increases sharply and it is not worth pursuing the snow-plow approximation to higher orders.

6.4 Explosion of a Finite Spherical Charge

A self-similar solution for the strong shock solution from the explosion of a finite spherical charge of radius R_0 cannot be obtained because of the presence of a characteristic length scale. However, an approximate solution can be determined using the snow-plow approximation.

Consider an explosive charge of radius R_0 with a large amount of energy E_0 released in the charge at time $t = 0$ generating a high-pressure P_0. We may write

$$E_0 = \left(\frac{k_j R_0^{j+1}}{j+1}\right)\left(\frac{P_0}{\gamma - 1}\right). \tag{6.4.1}$$

The first bracketed term on the right hand side denotes the charge volume and the second bracketed term represents the internal energy per unit volume. We shall neglect the initial internal energy of the charge. From Eq 6.4.1, the pressure of the charge after to the energy release is obtained as

$$P_0 = \frac{(j+1)(\gamma-1)E_0}{k_j R_0^{j+1}}. \tag{6.4.2}$$

Subsequent expansion of the charge boundary drives a strong shock wave outward and for $\gamma \simeq 1$, the mass swept by the shock is concentrated near the shock front. Thus we may write

$$R_c = R_s + \varepsilon R^{(1)} + \varepsilon^2 R^{(2)} + \cdots$$

where R_c is the charge radius. With $\varepsilon \ll 1$, $R_c \simeq R_s$ for the lowest order.

We shall assume the temperature (hence sound speed) of the gas inside the charge to be very high. Thus the pressure can equalize rapidly inside the charge and we may assume the pressure inside the charge to be uniform at all times. Also assuming the expansion of the gas to be isentropic, that is, pv^γ = constant, the pressure of the charge at any instant can be written as

$$\frac{p_c}{P_0} = \left(\frac{R_0}{R_c}\right)^{\gamma(j+1)}, \tag{6.4.3}$$

where $R_c(t)$ is the charge radius at any instant of time t. For $\gamma \simeq 1$, hence $\gamma - 1 = \delta \ll 1$, we write $\gamma = 1 + \delta$ and the above equation becomes

$$\frac{p_c}{P_0} = \left(\frac{R_0}{R_c}\right)^{(j+1)(1+\delta)} = \left(\frac{R_0}{R_c}\right)^{j+1} \left(\frac{R_0}{R_c}\right)^{\delta(j+1)} = \left(\frac{R_0}{R_c}\right)^{j+1} \left(e^{\ln\left(\frac{R_0}{R_c}\right)^{\delta(j+1)}}\right).$$

Expanding the exponential, we get

$$\frac{p_c}{P_0} = \left(\frac{R_0}{R_c}\right)^{j+1} \left(1 + \delta \ln \left(\frac{R_0}{R_c}\right)^{j+1} + O\left[\delta^2\right]\right).$$

For $\delta \ll 1$, $\frac{R_0}{R_c} \simeq O[1]$,

$$\frac{1}{\delta} \gg \ln \left(\frac{R_0}{R_c}\right)^{j+1},$$

or

$$e^{\frac{1}{\delta}} \gg \left(\frac{R_0}{R_c}\right)^{j+1}. \tag{6.4.4}$$

With $\delta \ll 1$, the above condition can readily be met since $\frac{R_0}{R_s} \simeq O[1]$.

The pressure can be written as a power series in ε, that is,

$$p = p^{(0)} + \varepsilon p^{(1)} + \cdots$$

and the solution for $p^{(0)}$ is given by Eq 6.3.9 as

$$p^{(0)} = \rho_0 \dot{R}_s^2 \left(1 + \frac{\ddot{R}_s R_s}{\dot{R}_s^2 (j+1)} \left(1 - \left(\frac{R}{R_s}\right)^{j+1}\right)\right).$$

At the interface (charge boundary) $R = 0$ and $p_c = p^{(0)}$, thus

$$p_c = P_0 \left(\frac{R_0}{R_c}\right)^{j+1} = \rho_0 \dot{R}_s^2 \left(1 + \frac{R_s \ddot{R}_s}{\dot{R}_s^2 (j+1)}\right),$$

where P_0 is given by Eq 6.4.2. The above equation can be written as

$$\frac{(j+1)(\gamma-1)E_0}{k_j\rho_0 R_s^{j+1}} = \dot{R}_s^2 + \frac{R_s\ddot{R}_s}{(j+1)}, \tag{6.4.5}$$

where we have replaced R_c by R_s as a first approximation. Defining $\phi(R_s) = \dot{R}_s^2$, we note $\ddot{R}_s = \frac{1}{2}\frac{d\phi}{dR_s}$ and we write Eq 6.4.5 as

$$\frac{A}{R_s^{j+1}} = \phi(R_s) + \frac{R_s}{2(j+1)}\frac{d\phi}{dR_s},$$

where $A = \frac{(j+1)(\gamma-1)E_0}{\rho_0 k_j}$. Multiplying throughout by $2(j+1)R_s^{2(j+1)-1}$, the above equation can be arranged as

$$\frac{d}{dR_s}\left(\phi R_s^{2(j+1)}\right) = 2(j+1)AR_s^j,$$

which can be integrated to yield

$$\phi R_s^{2(j+1)} = 2AR_s^{j+1} + B. \tag{6.4.6}$$

To evaluate the constant B, we can use the initial condition at $t = 0$ when $R_s = R_0$ and the shock pressure $p_1 = \rho_0\dot{R}_s^2 = P_0$. Thus $\dot{R}_s^2 = \phi(R_0) = \frac{P_0}{\rho_0}$ and Eq 6.4.2 gives

$$\dot{R}_s^2(R_s = R_0) = \frac{P_0}{\rho_0} = \frac{(j+1)(\gamma-1)E_0}{\rho_0 k_j R_0^{j+1}} = \frac{A}{R_0^{j+1}}.$$

The integration constant is found to be

$$B = -AR_0^{j+1},$$

and Eq 6.4.6 becomes

$$\dot{R}_s^2 = \frac{2A}{R_s^{j+1}}\left(1 - \frac{1}{2}\left(\frac{R_0}{R_s}\right)^{j+1}\right), \tag{6.4.7}$$

or

$$\dot{R}_s^2 = \left[\frac{2(j+1)(\gamma-1)E_0}{\rho_0 k_j}\right]\frac{1}{R_s^{j+1}}\left(1 - \frac{1}{2}\left(\frac{R_0}{R_s}\right)^{j+1}\right). \tag{6.4.8}$$

When $R_s \gg R_0$, the above equation reduces to

$$\dot{R}_s^2 = \left[\frac{2(j+1)(\gamma-1)E_0}{\rho_0 k_j}\right]\frac{1}{R_s^{j+1}},$$

which corresponds to the point source constant energy solution. Since $\gamma \simeq 1$, the zeroth-order solution for the density and particle velocity distributions can be obtained in a manner similar to the previous case of the point source solution considered in Section 6.3.

The present result demonstrates the stability of the point source solution. When the initial condition deviates from the point source solution, the point source solution is recovered asymptotically at large radius $R_s \gg R_0$ where the non-similar effect due to the presence of a characteristic length scale R_0 diminishes.

6.5 Piston Driven Explosions

In the explosion of a finite charge, the expanding interface essentially acts like a piston driving the shock ahead of it. However, the piston path is coupled to the blast motion in that the pressure is uniform across the expanding interface. In this section, the motion of a solid piston with a prescribed path driving the shock is considered. Thus the piston motion does not have to be solved simultaneously with the blast motion. It is now a prescribed boundary condition. Piston driven explosions are analogous to the hypersonic flow over blunted nosed bodies under the hypersonic similitude rule. The snow-plow approximation is equivalent to the Newtonian flow approximation in hypersonic flow which has been investigated by Cole (1961), Freeman (1958), and Chernyi (1961) among others.

Since the piston path $R_p(t)$ is in essence a particle path $R =$ constant, we may write

$$r(R_p, t) = R_p(t) = R_s + \varepsilon R^{(1)} + \varepsilon^2 R^{(2)} + \cdots$$

and the solution for $R^{(1)}$ is given by Eq 6.2.17, that is,

$$R^{(1)} = -\int_R^{R_s} \left[\frac{\dot{R}_1^2(\tau)}{\dot{R}_s^2(t)}\right]^{\frac{1}{\gamma}} \frac{\left(\frac{R}{R_s}\right)^j \mathrm{d}R}{\left[1 + \frac{R_s\ddot{R}_s}{\dot{R}_s^2(j+1)}\left(1 - \left(\frac{R}{R_s}\right)^{j+1}\right)\right]^{\frac{1}{\gamma}}},$$

where τ is the time where the particle R first crosses the shock. If the shock path $R_s(t)$ is specified and the corresponding piston path is to be found (the inverse problem), the above equations can be integrated. However, if the piston path $R_p(t)$ is specified and the shock trajectory is to be determined, the problem is more involved. Replacing $R_s(t)$ by $R_p(t)$ in above equations will involve an error of the order of ε^2 only.

Quite often it is more convenient to use the time τ at which a particle enters the shock instead of R as the Lagrangian variable. τ is the time when the particle first crosses the shock front, that is, $R = R_s(\tau)$ as shown in Fig. 6.4.

Since $dR = \dot{R}(\tau)\mathrm{d}\tau$, Eq 6.2.17 can be transformed to read

$$R^{(1)} = -\int_\tau^t \frac{\dot{R}_s^{\frac{2}{\gamma}+1} R_s^j(\tau)\mathrm{d}\tau}{R_s^j(t)\left[\dot{R}_s^2(t) + \frac{R_s\ddot{R}_s}{\dot{R}_s^2(j+1)}\left(1 - \left(\frac{R_s(\tau)}{R_s(t)}\right)^{j+1}\right)\right]^{\frac{1}{\gamma}}}. \quad (6.5.1)$$

Thus if $R_s(\tau)$ is specified, the above equation can be integrated. Alternately, if $R_p(t)$ is specified, we may replace R_s by R_p in Eq 6.5.1 and write

$$R^{(1)} = -\int_\tau^t \frac{\dot{R}_p^{\frac{2}{\gamma}+1} R_p^j(\tau)\mathrm{d}\tau}{R_p^j(t)\left[\dot{R}_p^2(t) + \frac{R_p\ddot{R}_p}{\dot{R}_p^2(j+1)}\left(1 - \left(\frac{R_p(\tau)}{R_p(t)}\right)^{j+1}\right)\right]^{\frac{1}{\gamma}}}. \quad (6.5.2)$$

In general, the integration of Eqs 6.5.1 and 6.5.2 have to be carried out numerically. First-order solutions for the pressure and density can be obtained from Eqs 6.2.18 and 6.2.19.

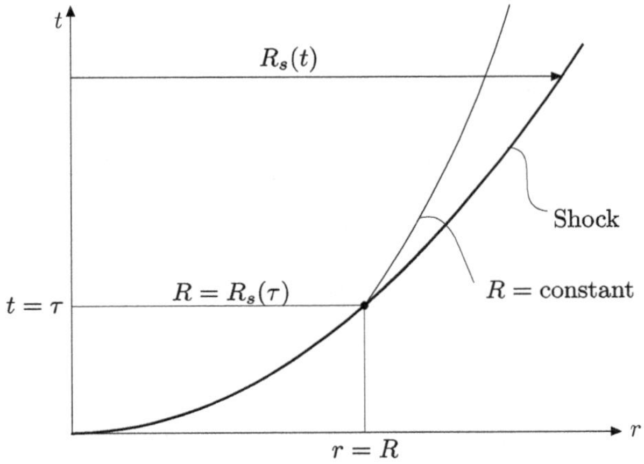

Figure 6.4 Particle Entering the Shock at a Given Time

If we specify the piston velocity instead, that is, $R_p = D't$ where D' is the piston velocity, then Eq 6.5.2 gives

$$R^{(1)} = \frac{-D't}{(j+1)}\left(1-\left(\frac{\tau}{t}\right)^{j+1}\right),$$

and the shock path is then

$$R_s(t) = R_p(t) - \varepsilon R^{(1)} = \left(\frac{2\gamma + j(\gamma+1)}{(j+1)(\gamma+1)}\right) D't.$$

For the planar case where $j = 0$, we obtain

$$R_s(t) = \frac{2\gamma}{(\gamma+1)} D't,$$

or the piston velocity

$$\frac{u_p}{\dot{R}_s} = \frac{D'}{\dot{R}_s} = \frac{\gamma+1}{2\gamma}. \tag{6.5.3}$$

The piston velocity should be the same as the particle velocity behind the strong shock, that is,

$$\frac{u_1}{\dot{R}_s} = \frac{2}{\gamma+1}.$$

However, for $\gamma \simeq 1$, the two expressions do not differ significantly. For example, from Eq 6.5.3

$$\dot{R}_s = \frac{2\gamma}{\gamma+1} D',$$

and if $(\gamma - 1) = \delta$ (or $\gamma = 1 + \delta$), the above equation becomes

$$\dot{R}_s = \frac{2(1+\delta)}{2+\delta} D' = (1+\delta)\left(1-\frac{\delta}{2}+\cdots\right) D' = \left(1+\frac{\delta}{2}+\cdots\right) D'.$$

Taking first order in δ only

$$\frac{\dot{R}_s}{D'} = \left(1 + \frac{\gamma - 1}{2}\right) = \frac{\gamma + 1}{2},$$

which is just the strong shock condition $\frac{\dot{R}_s}{u_1} = \frac{\gamma+1}{2}$. Thus the results differ by terms of the order of δ^2.

For the pressure distribution, we note from Eq 6.2.15 that

$$p^{(0)} = \rho_0 \dot{R}_s^2,$$

since $\ddot{R}_s = 0$ for the case of a constant velocity shock. For the first-order solution $p^{(1)}$, Eq 6.2.18 gives

$$p^{(1)} = -\rho_0 \dot{R}_s^2 + \int_R^{R_s} \left[\rho_0 \frac{\partial^2 R^{(1)}}{\partial t^2} \left(\frac{R}{R_s}\right)^j - \frac{j\rho_0 \ddot{R}_s R^j R^{(1)}}{R_s^{j+1}}\right] \mathrm{d}R,$$

and since $\ddot{R}_s = 0$, the second term of the integrand vanishes. With $R^{(1)}$ given by Eq 6.5.3, we obtain

$$\frac{\partial^2 R^{(1)}}{\partial t^2} = \frac{jD}{\tau} \left(\frac{\tau}{r}\right)^{j+2},$$

and the solution for $p^{(1)}$ is written as

$$p^{(1)} = -\rho_0 D^2 + \int_\tau^t \rho_0 \left(\frac{jD}{\tau} \left(\frac{\tau}{t}\right)^{j+2}\right) \left(\frac{D\tau}{Dt}\right)^j \mathrm{d}(D\tau),$$

where we have replaced $R = D\tau$, $R_s = Dt$ and $\dot{R}_s = D$ and changed the integration with respect to R to time τ.

Evaluating the integral, we get

$$p^{(1)} = -\rho_0 D^2 + \frac{\rho_0 D^2 j}{2(j+1)} \left(1 - \left(\frac{\tau}{t}\right)^{2(j+1)}\right). \tag{6.5.4}$$

Thus the pressure can be written as

$$\begin{aligned} p &= p^{(0)} + \varepsilon p^{(1)} + \cdots \\ &= \rho_0 D^2 - \left(\frac{\gamma - 1}{\gamma + 1}\right) \rho_0 D^2 \left(1 - \frac{j}{2(j+1)} \left(1 - \left(\frac{\tau}{t}\right)^{j+1}\right)\right) + \cdots \end{aligned} \tag{6.5.5}$$

Equation 6.2.10 can be integrated to give the density distribution as

$$\left(\frac{p^{(1)}}{p^{(0)}} - \frac{\gamma \rho^{(1)}}{\rho^{(0)}}\right) = \text{function of } R.$$

Evaluating the right hand side at the shock front $R = R_s$ where $p_1^{(1)} = -\rho_0 \dot{R}_s^2$, and $p_1^{(0)} = -\rho_0 \dot{R}_s^2$, $\rho_1^{(0)} = -\rho_0$, $\rho_1^{(1)} = 0$, we get

$$\rho^{(1)} = \frac{\rho^{(0)}}{\gamma} \left(1 + \frac{p^{(1)}}{p^{(0)}}\right).$$

Equation 6.2.15 gives $p^{(0)} = \rho_0 \dot{R}_s^2$ and Eq 6.2.16 gives $\rho^{(0)} = \rho_0$ (since $\dot{R}_s(t) = \dot{R}_s(\tau) = D$ and $\ddot{R}_s = 0$) and using Eq 6.5.4 for $p^{(1)}$, we get

$$\rho^{(1)} = \frac{\rho_0 j}{2\gamma(j+1)}\left(1 - \left(\frac{\tau}{t}\right)^{2(j+1)}\right). \tag{6.5.6}$$

Thus the density distribution can be written as

$$\begin{aligned}\rho &= \frac{\rho^{(0)}}{\varepsilon} + \rho^{(1)} + \cdots \\ &= \left(\frac{\gamma+1}{\gamma-1}\right)\rho_0 + \frac{\rho_0 j}{2\gamma(j+1)}\left(1 - \left(\frac{\tau}{t}\right)^{2(j+1)}\right).\end{aligned} \tag{6.5.7}$$

With the solution for $R^{(1)}$ the particle path can be obtained from

$$\begin{aligned}r(R,t) &= R_s(t) + \varepsilon R^{(1)} \\ &= Dt\left[1 + \left(\frac{\gamma-1}{\gamma+1}\right)\frac{1}{(j+1)}\left(1 - \left(\frac{\tau}{t}\right)^{j+1}\right)\right].\end{aligned} \tag{6.5.8}$$

The particle velocity is given by

$$u = \left(\frac{\partial r}{\partial t}\right)_R = \dot{R}_s + \varepsilon\frac{\partial R^{(1)}}{\partial t} + \cdots$$

Differentiating $R^{(1)}$ with respect to t, the above expression becomes

$$\frac{u}{D} = \left\{\frac{(\gamma+1)(j+1) + (\gamma-1)}{(\gamma+1)(j+1)} + \frac{j(\gamma-1)}{(\gamma+1)(j+1)}\left(\frac{\tau}{t}\right)^{j+1}\right\}, \tag{6.5.9}$$

where $\tau = 0$ corresponds to the piston path. To obtain the velocity profile in terms of the spatial distribution r, we solve for $\frac{\tau}{t}$ in terms of $\frac{r}{R_s}$, that is,

$$\begin{aligned}\frac{r}{R_s} &= 1 + \varepsilon\frac{R^{(1)}}{R_s} + \cdots \\ &= 1 + \varepsilon\left(\frac{1}{j+1}\right)\left(1 - \left(\frac{\tau}{t}\right)^{j+1}\right).\end{aligned}$$

Solving the above equation yields

$$\left(\frac{\tau}{t}\right)^{j+1} = \frac{(\gamma-1) + (j+1)(\gamma+1)}{(\gamma-1)} - \frac{(j+1)(\gamma+1)}{(\gamma-1)}\left(\frac{r}{R_s}\right), \tag{6.5.10}$$

and we can replace $\left(\frac{\tau}{t}\right)$ in Eq 6.5.9 by the above equation.

For more general piston motions, the integral for $R^{(1)}$ (e.g., Eqs 6.5.9, 6.5.10) will be more complicated and requires numerical integration. For a power law shock path $R_s = At^N$, the shock velocity and acceleration will be given by $\dot{R}_s = ANt^{N-1}$, $\ddot{R}_s = AN(N-1)t^{N-2}$ and $\frac{R_s\ddot{R}_s}{\dot{R}_s^2} = \frac{N-1}{N} = \theta$. When N is specified, θ is known and the zeroth-order pressure given by Eq 6.2.15 becomes

$$p^{(0)} = \rho_0 \dot{R}_s^2\left[1 + \frac{\theta}{(j+1)}\left(1 - \left(\frac{R}{R_s}\right)^{j+1}\right)\right]. \tag{6.5.11}$$

If τ is used as the Lagrangian variable instead of R, that is, $R = R_s(\tau)$, then $p^{(0)}$ can be written as

$$p^{(0)} = \rho_0 \dot{R}_s^2 \left[1 + \frac{\theta}{(j+1)}\left(1 - \left(\frac{\tau}{t}\right)^{N(j+1)}\right)\right], \tag{6.5.12}$$

since $R = R_s(\tau) = A\tau^N$. The density $\rho^{(0)}$ from Eq 6.5.12 is written as

$$\rho^{(0)} = \rho_0 \left\{\left[\frac{1}{\left(\frac{\tau}{t}\right)^{2(N-1)}}\right]\left[1 + \frac{\theta}{j+1}\left(1 - \left(\frac{\tau}{t}\right)^{N(j+1)}\right)\right]\right\}, \tag{6.5.13}$$

The solution for $R^{(1)}$ given by Eq 6.2.17 becomes

$$R^{(1)} = \frac{-AN}{t^{\frac{2}{\gamma}(N-1)+jN}} \int_\tau^t \frac{\tau^{(\frac{2}{\gamma}+1)(N-1)+jN} \mathrm{d}\tau}{\left[1 + \frac{\theta}{j+1}\left(1 - \left(\frac{\tau}{t}\right)^{N(j+1)}\right)\right]^{\frac{1}{\gamma}}}. \tag{6.5.14}$$

Given N (hence θ), the above integral has to be evaluated numerically. With $p^{(0)}$, $\rho^{(0)}$, and $R^{(1)}$ known, first-order solutions for $p^{(1)}$ and $\rho^{(1)}$ can be obtained accordingly. The piston path is given by

$$R_p(t) = R_s(t) + \varepsilon R^{(1)}(\tau = 0) + \cdots$$

As can be observed, the solution for general piston motion is rather complex and, in a way, renders the simplicity of the snow-plow approximation less attractive.

7 The Brinkley–Kirkwood Theory

7.1 Introduction

The description of the propagation of non-steady shock waves requires the integration of the partial differential equations for the flow behind the shock subject to the boundary conditions at the shock and at some rear boundary. Since the shock velocity is not known, the numerical solution is not straight forward. There are practical problems where a complete description is not required, and only the variation of the shock strength with position is desired. Thus there is a need for approximate methods of solution for the shock front only. During the early 1940s, Kirkwood and Brinkley (1945) and Brinkley and Kirkwood (1947) developed a simple method to describe blast waves in water and air. They derived a simple "shock front evolution equation" which gave surprisingly good predictions for the blast wave decay in air and water. Brinkley and Kirkwood (henceforth referred to as BK) derived a pair of ordinary differential equations for the variation of the shock strength and the blast energy with shock radius from the conservation equations, the Rankine–Hugoniot conditions at the shock front, and the energy integral. The unique feature of the BK theory is the formulation of the blast energy integral. The blast energy is defined as the sum of the kinetic and internal energy between the shock front and a particle path which acts like a piston. The blast energy is then equated to the work done by the particle path. Since no mass crosses a particle trajectory, it is essentially a solid expanding piston that displaces the fluid ahead of it. When time is taken to infinity where all motion ceases, the total work done is manifested as the residual internal energy in the shocked fluid. The residual internal energy also corresponds to the energy dissipated by the shock wave. The BK theory recognizes the finite entropy increment of the fluid particle that crosses the shock wave. The energy dissipation reduces the available energy of the shock wave and results in its decay. Thus the BK theory addresses explicitly the mechanism of shock decay.

Descriptions of the BK method, described in the wartime NDRC reports and in the Physical Review paper by Brinkley and Kirkwood, are rather brief and difficult to follow. The reviews of the BK theory by Baker (1974) and Cole (1961) are equally simplified, aiming only at giving a brief introduction of the essential features of the theory. However, a more detailed development of the BK theory was given by Bach, Chiu, and Lee (1975) and Chiu, Lee, and Knystautas (1977). The analysis in the present chapter follows closely the work of Bach et al. (1975).

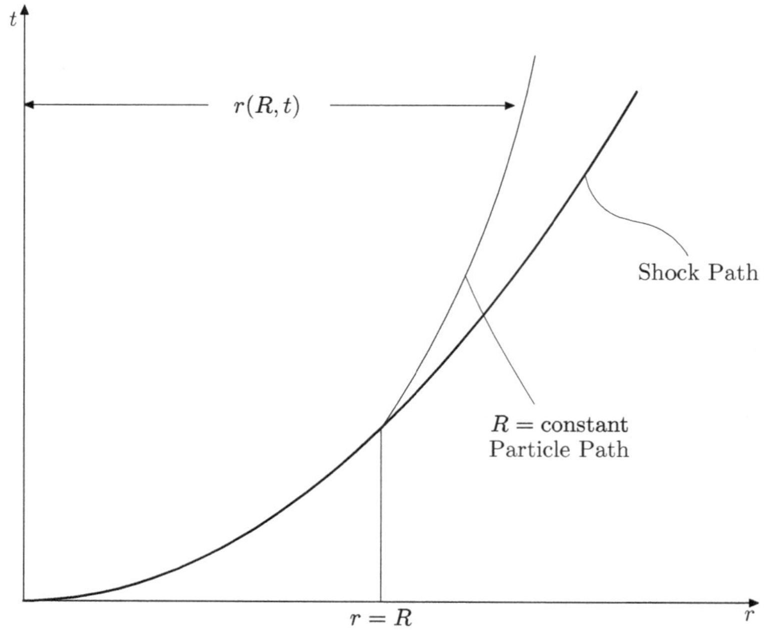

Figure 7.1 Shock Trajectory and Particle Path Behind the Shock

7.2 Basic Equations

The conservation equations for one-dimensional non-steady flow with planar ($j = 0$), cylindrical ($j = 1$), and spherical ($j = 2$) symmetry in Lagrangian form are given as

$$\left(\frac{\partial r}{\partial R}\right)_t = \left(\frac{\rho_0}{\rho}\right)\left(\frac{R}{r}\right)^j, \tag{7.2.1}$$

$$\left(\frac{\partial u}{\partial t}\right)_R = -\frac{1}{\rho_0}\left(\frac{r}{R}\right)^j\left(\frac{\partial p}{\partial R}\right)_t, \tag{7.2.2}$$

$$\left(\frac{\partial s}{\partial t}\right)_R = \frac{\partial}{\partial t}\left(\frac{p}{\rho^\gamma}\right)_R, \tag{7.2.3}$$

where R is the Lagrangian variable and chosen to correspond to the initial position of a fluid particle. At a later time, the particle R will be at $r(R, t)$. Figure 7.1 shows the motion of a particle behind a shock. The dependent variables $p(R, t)$, $\rho(R, t)$, $u(R, t) = \left(\frac{\partial r}{\partial t}\right)_R$ have their usual meaning. An alternate form of the continuity equation used by Brinkley and Kirkwood is obtained by differentiating Eq 7.2.1 with respect to time, that is,

$$\frac{1}{\rho}\left(\frac{\partial \rho}{\partial t}\right)_R + \left(\frac{r}{R}\right)^j\left(\frac{\rho}{\rho_0}\right)\left(\frac{\partial u}{\partial R}\right)_t + \frac{ju}{r} = 0. \tag{7.2.4}$$

With $s(p, \rho)$, we can write

$$\frac{\partial s}{\partial t} = \left(\frac{\partial s}{\partial \rho}\right)_p \frac{\partial \rho}{\partial t} + \left(\frac{\partial s}{\partial p}\right)_\rho \frac{\partial p}{\partial t} = 0,$$

and thus

$$\frac{\partial \rho}{\partial t} = \frac{-\left(\frac{\partial s}{\partial p}\right)_\rho}{\left(\frac{\partial s}{\partial \rho}\right)_p} \frac{\partial p}{\partial t},$$

and from the cyclic rule

$$\left(\frac{\partial s}{\partial p}\right)_\rho \left(\frac{\partial p}{\partial \rho}\right)_s \left(\frac{\partial \rho}{\partial s}\right)_p = -1,$$

we get

$$\frac{\partial \rho}{\partial t} = \left(\frac{\partial \rho}{\partial p}\right)_s \frac{\partial p}{\partial t} = \frac{1}{c^2}\frac{\partial p}{\partial t}.$$

Equation 7.2.4 can then be written as

$$\frac{1}{\rho c^2}\left(\frac{\partial p}{\partial t}\right)_R + \frac{ju}{r} + \left(\frac{r}{R}\right)^j \left(\frac{\rho}{\rho_0}\right)\left(\frac{\partial u}{\partial R}\right)_t = 0. \tag{7.2.5}$$

Equations 7.2.2, 7.2.3, and 7.2.5 are valid for any particle R which corresponds to the shock position when the particle first crosses the shock.

If we evaluate Eqs 7.2.1 and 7.2.5 at the shock front $r = R$, we get

$$\left(\frac{\partial u_s}{\partial t}\right) + \frac{1}{\rho_0}\frac{\partial p_s}{\partial R} = 0, \tag{7.2.6}$$

$$\frac{1}{\rho_s c_s^2}\frac{\partial p_s}{\partial t} + \frac{ju_s}{R} + \left(\frac{\rho_s}{\rho_0}\right)\left(\frac{\partial u_s}{\partial R}\right) = 0, \tag{7.2.7}$$

where the subscript s denotes to the condition at the shock front. Equations 7.2.6 and 7.2.7 give the variations of p_s and u_s with time and shock position R. Note that the shock position is now defined by the position of the particle R that first crosses the shock.

To complete the formulation, we need to derive an equation for p_s (or u_s) along the shock path of the form $\frac{\mathrm{d}p_s}{\mathrm{d}R}$ (or $\frac{\mathrm{d}u_s}{\mathrm{d}R}$) where $\frac{\mathrm{d}}{\mathrm{d}R}$ is the variation along the shock trajectory. For any function $F(R, t)$,

$$\frac{\mathrm{d}F}{\mathrm{d}R} = \frac{\partial F}{\partial R} + \frac{\partial F}{\partial t}\frac{\partial t}{\partial R},$$

and evaluating at the shock, we write

$$\frac{\mathrm{d}F}{\mathrm{d}R} = \frac{\partial F}{\partial R} + \frac{1}{D}\frac{\partial F}{\partial t},$$

where $D = \frac{\mathrm{d}R}{\mathrm{d}t}$ is the shock velocity. Equations 7.2.6 and 7.2.7 give two equations for the four derivatives $\frac{\partial p_s}{\partial t}$, $\frac{\partial p_s}{\partial R}$, $\frac{\partial u_s}{\partial t}$ and $\frac{\partial u_s}{\partial R}$. We need two more equations to solve for all the derivatives. The Rankine–Hugoniot equation across the shock gives

$$p_s - p_0 = \Delta p_s = \rho_0 \dot{R} u_s = \rho_0 D u_s \tag{7.2.8}$$

Thus

$$\frac{\mathrm{d}\Delta p_s}{\mathrm{d}R} = \rho_0 D \frac{\mathrm{d}u_s}{\mathrm{d}R} + \rho_0 u_s \frac{\mathrm{d}D}{\mathrm{d}R}, \tag{7.2.9}$$

where $\frac{\mathrm{d}}{\mathrm{d}R} = \frac{\partial}{\partial R} + \frac{1}{D}\frac{\partial}{\partial t}$ is the total derivative along the shock path. The Rankine–Hugoniot relationships also give

$$D^2 = c_0^2 + \frac{\gamma+1}{2}\frac{\Delta p_s}{\rho_0}, \tag{7.2.10}$$

from which we obtain

$$\frac{\mathrm{d}D}{\mathrm{d}R} = \frac{\gamma+1}{4\rho_0 D}\frac{\mathrm{d}\Delta p_s}{\mathrm{d}R}. \tag{7.2.11}$$

Substituting the above equation into Eq 7.2.9 gives a relationship between $\frac{\mathrm{d}\Delta p_s}{\mathrm{d}R}$ and $\frac{\mathrm{d}u_s}{\mathrm{d}R}$ as

$$\frac{\mathrm{d}u_s}{\mathrm{d}R} = \frac{1}{\rho_0 D}\left[1 - \frac{\gamma+1}{4}\frac{u_s}{D}\right]\frac{\mathrm{d}\Delta p_s}{\mathrm{d}R}. \tag{7.2.12}$$

The Rankine–Hugoniot relationships also give

$$\frac{u_s}{D} = \frac{2\frac{\Delta p_s}{p_0}}{2 + (\gamma+1)\frac{\Delta p_s}{p_0}}. \tag{7.2.13}$$

Eqs 7.2.6, 7.2.7, and 7.2.12 now provide us with three equations for the four derivatives.

7.3 The Energy Integral

The remaining fourth equation is obtained from the conservation of energy. Brinkley and Kirkwood considered a novel form of the energy integral by equating the work done by a particle path to the gain in the internal and kinetic energies of the shocked fluid between the shock front and the particle path, that is,

$$\begin{aligned} W(R_0, t) &= \int_{t_0(R_0)}^{t} k_j r^j p(R_0, t) u(R_0, t)\mathrm{d}t, \\ &= \int_{r(R_0,t)}^{R} k_j r^j (R, t)\mathrm{d}r \rho(R, t)\left(\Delta e + \frac{u^2}{2}\right), \end{aligned} \tag{7.3.1}$$

where $t_0(R_0)$ is the time the shock arrives at $r = R_0$, $r(R_0, t)$ is the position of the particle R_0 at time t.

The first integral on the right hand side of Eq 7.3.1 is carried out along the path $R_0 =$ constant whereas the second integral is evaluated at constant time t between the

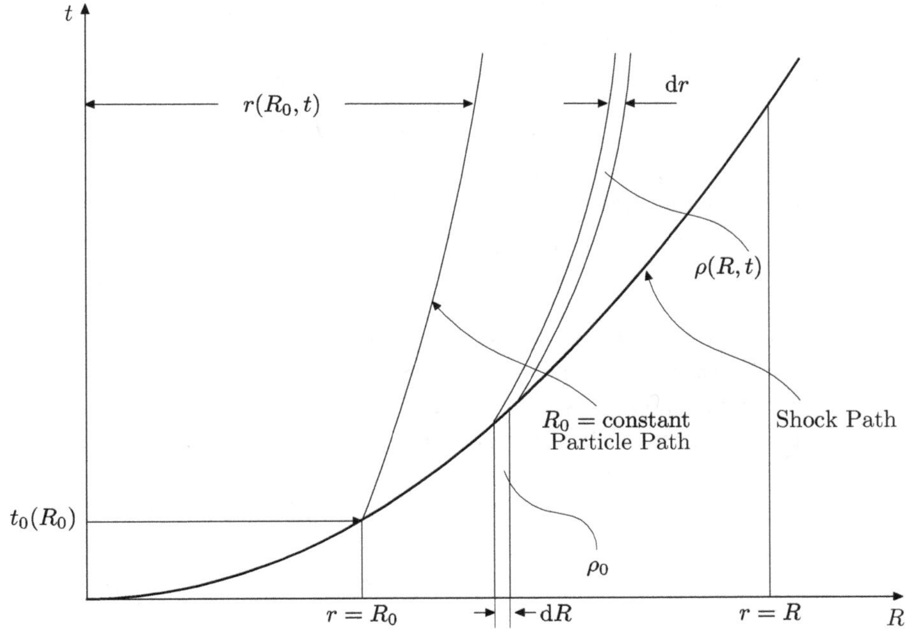

Figure 7.2 Schematic of Work Done by a Particle Path

shock $r = R$ and the particle path $r = r(R_0, t)$. This is illustrated in Fig. 7.2. Using the conservation of mass, that is, $k_j r^j \rho \mathrm{d}r = k_j R^j \rho_0 \mathrm{d}R$, we can write the second integral as

$$\int_{R_0}^{R} k_j R^j \rho_0 \left(\Delta e + \frac{u^2}{2} \right) \mathrm{d}R.$$

If we let $t \to \infty$, the total work done by the particle R_0 is given by

$$\begin{aligned} W(R_0, \infty) &= \int_{t_0(R_0)}^{\infty} k_j r^j p(R,t) u(R,t) \mathrm{d}t, \\ &= \int_{R_0}^{\infty} k_j r^j \rho_0 \Delta e_\infty \mathrm{d}R, \end{aligned} \tag{7.3.2}$$

where $\Delta e_\infty = e_\infty - e_0$ is the residual internal energy of the shocked fluid. When $t \to \infty$, $u \to 0$, thus the kinetic energy term vanishes. We may also drop the subscript 0 and write $W(R, \infty)$ as $W(R)$.

Brinkley and Kirkwood considered the overpressure work rather than the pressure work. Since $\Delta p = p - p_0$, we write

$$\begin{aligned} W(R) &= \int_{t}^{\infty} k_j r^j p(R,t) u(R,t) \mathrm{d}t \\ &= \int_{t}^{\infty} k_j r^j \Delta p u \mathrm{d}t + \int_{t}^{\infty} k_j r^j p_0 u \mathrm{d}t. \end{aligned} \tag{7.3.3}$$

Since $u = \left(\frac{\partial u}{\partial t}\right)_R$ and the second integral on the right hand side of the above is taken along a particle path R = constant we can write

$$\int_t^\infty k_j r^j p_0 \frac{\mathrm{d}u}{\mathrm{d}t}\mathrm{d}t = \int_{r=R}^{r(R,\infty)} k_j r^j p_0 \mathrm{d}r,$$

where $r(R, \infty)$ is the eventual position of the particle R when $t \to \infty$. We may write the integral on the right hand side of the above as follows:

$$\int_{r=R}^{r(R,\infty)} k_j p_0 r^j \mathrm{d}r = \int_{r=R}^{\infty} p_0 k_j r^j \mathrm{d}r - \int_{r(R,\infty)}^{\infty} k_j p_0 r^j \mathrm{d}r,$$

where r in the first integral on the right hand side of the above equation is a dummy integration variable and can be replaced by R and the second integral can also be changed to an R integration using the conservation of mass $k_j r^j \mathrm{d}r \rho_\infty = k_j R^j \mathrm{d}R \rho_0$. This is because the integration is taken at constant $t \to \infty$. Thus

$$\begin{aligned}\int_{r=R}^{r(R,\infty)} k_j p_0 r^j \mathrm{d}r &= \int_R^\infty p_0 k_j R^j \mathrm{d}R - \int_R^\infty k_j p_0 \frac{\rho_0}{\rho_\infty} R^j \mathrm{d}R,\\ &= \int_R^\infty k_j p_0 R^j \rho_0 \left(\frac{1}{\rho_0} - \frac{1}{\rho_\infty}\right)\mathrm{d}R.\end{aligned}$$

Eqs 7.3.2 and 7.3.3 now become

$$\begin{aligned}W(R) &= \int_R^\infty k_j R^j \rho_0 \Delta e_\infty \mathrm{d}R\\ &= E(R) + \int_R^\infty k_j R^j \rho_0 p_0 \left(\frac{1}{\rho_0} - \frac{1}{\rho_\infty}\right)\mathrm{d}R, \qquad (7.3.4)\end{aligned}$$

where $E(R)$ is the overpressure work given by

$$E(R) = \int_t^\infty k_j r^j \Delta p(R,t) u(R,t)\mathrm{d}t. \qquad (7.3.5)$$

From Eq 7.3.4, we obtain

$$\begin{aligned}E(R) &= \int_R^\infty k_j R^j \rho_0 \left(e_\infty - e_0 - \frac{p}{\rho_0} + \frac{p_\infty}{\rho_\infty}\right)\mathrm{d}R,\\ &= \int_R^\infty k_j R^j \rho_0 (h_\infty - h_0)\mathrm{d}R, \qquad (7.3.6)\end{aligned}$$

where $h = e + \frac{p}{\rho}$ is the enthalpy and we have used the fact that as $t \to \infty$, $p_\infty \to p_0$. Equation 7.3.6 states that the overpressure work done by a particle eventually becomes the residual enthalpy of the shocked fluid whereas the pressure work becomes the residual internal energy.

The residual internal energy and enthalpy can be expressed in terms of p and ρ since for a perfect gas $e = \frac{p}{\rho(\gamma-1)}$ and $h = \frac{\gamma p}{\rho(\gamma-1)}$. Thus we write

$$h_\infty - h_0 = h_0\left(\frac{h_\infty}{h_0} - 1\right) = \frac{\gamma p_0}{\rho_0(\gamma-1)}\left(\frac{\rho_0}{\rho_\infty} - 1\right),$$

since $p_\infty = p_0$. Using the particle isentropic condition, $\left(\frac{p}{\rho^\gamma}\right)_R = \text{constant}$, we may write $\frac{p_s}{\rho_s^\gamma} = \frac{p_\infty}{\rho_\infty^\gamma}$ and hence $\frac{\rho_0}{\rho_\infty} = \left(\frac{p_s}{p_0}\right)^{\frac{1}{\gamma}}\left(\frac{\rho_0}{\rho_s}\right) = \left(\frac{\Delta p_s}{p_0}+1\right)^{\frac{1}{\gamma}}\left(\frac{\rho_0}{\rho_s}\right)$ where we have used $p_\infty = p_0$. Thus we may write Eq 7.3.6 in terms of Δp_s and ρ_s as

$$E(R) = \int_R^\infty k_j R^j \mathrm{d}R \frac{\gamma p_0}{(\gamma-1)}\left[\left(\frac{\Delta p_s}{p_0}+1\right)^{\frac{1}{\gamma}}\frac{\rho_0}{\rho_s} - 1\right]. \tag{7.3.7}$$

Eq 7.3.7 above gives the variation of the overpressure work done by any particle R, or equivalently Eq 7.3.7 gives the variation of the overpressure work as a function of the shock position R. The right hand side of Eq 7.3.7 is the residual enthalpy. The energy of the blast wave comes from the work done by the particle and eventually the energy of the blast wave remains as residual internal energy or enthalpy. Thus the BK theory links directly the energy dissipation by the shock wave to its decay.

7.4 The Fourth Equation

The fourth equation for the derivatives can now be obtained from the energy integral. We can write Eq 7.3.5 as

$$\begin{aligned} E(R) &= k_j R^j \Delta p_s u_s \int_{t(R)}^\infty \left(\frac{r}{R}\right)^j \left(\frac{\Delta p}{\Delta p_s}\right)\left(\frac{u}{u_s}\right)\mathrm{d}t, \\ &= k_j R^j \Delta p_s u_s \int_{t(R)}^\infty f \mathrm{d}t, \end{aligned} \tag{7.4.1}$$

where

$$f = \left(\frac{r}{R}\right)^j \left(\frac{\Delta p}{\Delta p_s}\right)\left(\frac{u}{u_s}\right), \tag{7.4.2}$$

At the shock, $r = R$, $\Delta p = \Delta p_s$, $u = u_s$ and the integrand $f = 1$. As $t \to \infty$, $\Delta p \to 0$, $u \to 0$, $f \to 0$. Thus f is a decreasing function of time. Near the shock, we may approximate $f \simeq e^{-\tau}$ where

$$\tau = \frac{t - t(R)}{\mu}. \tag{7.4.3}$$

Thus $\mathrm{d}t = \mu(R)\mathrm{d}\tau$ where $\frac{1}{\mu} = -\frac{1}{f}\frac{\partial f}{\partial t}$. Equation 7.4.1 can be written as

$$E(R) = k_j R^j \Delta p_s u_s \mu(R)\nu(R), \tag{7.4.4}$$

where

$$\nu(R) = \int_0^\infty f \mathrm{d}\tau, \tag{7.4.5}$$

is the dimensionless integral (Eq 7.4.5). To evaluate $\nu(R)$, the variation of $f(\tau)$ (or $\Delta p(R,t)$ and $u(R,t)$) with time τ must be known. This amounts to knowing the solution for the flow field behind the shock. For $f(\tau) \sim e^{-\tau}$, the integral $\nu \simeq 1$. For the regime corresponding to a linear pressure–time profile of the positive phase of the wave, $f(\tau) = (1 - \tau/2)^2$ for $\tau \leq 2$ and $f(\tau) = 0$ for $\tau > 2$. This gives $\nu = \frac{2}{3}$. Thus $\frac{2}{3} \leq \nu \leq 1$ and the variation of ν is not large. Brinkley and Kirkwood pointed out that assigning a value for ν independent of R is equivalent to imposing a similarity constraint on the energy–time curve.

In the original BK analysis, they chose $\nu(R)$ to be of the form

$$\nu = 1 - \frac{1}{3} e^{-\frac{\Delta p_s}{p_0}}, \tag{7.4.6}$$

so that for strong shocks where $\frac{\Delta p_s}{p_0} \gg 1$, $\nu \to 1$ and for the weak shock regime where $\frac{\Delta p_s}{p_0} \ll 1$, $\nu \to \frac{2}{3}$. Thus Eq 7.4.4 recovers the two limits of strong and weak shocks. Assuming a form for the energy integral eliminates the need to obtain a solution of the basic equations. The advantage of the BK theory is to formulate the problem in such a manner so that the blast decay can be solved without integrating the partial differential equations of motion. The blast decay is not sensitive to the energy integral ν which is a slowly varying function between a small range of values.

From Eq 7.4.2, the fourth equation for the derivatives $\frac{\mathrm{d}\Delta p_s}{\mathrm{d}R}$, $\frac{\partial \Delta p_s}{\partial t}$, etc. can now be obtained. Differentiating Eq 7.4.2 with respect to time t along the particle path $R =$ constant yields

$$\frac{1}{f}\left(\frac{\partial f}{\partial t}\right)_R = \frac{j}{r}\left(\frac{\partial r}{\partial t}\right)_R + \frac{1}{\Delta p}\left(\frac{\partial \Delta p}{\partial t}\right)_R + \frac{1}{u}\left(\frac{\partial u}{\partial t}\right)_R,$$

and evaluating the above equation at the shock front gives

$$-\frac{1}{\mu(R)} = \frac{j u_s}{R} + \frac{1}{\Delta p_s}\frac{\partial \Delta p_s}{\partial t} + \frac{1}{u_s}\frac{\partial u_s}{\partial t}, \tag{7.4.7}$$

where we have used Eq 7.4.3 for $\frac{1}{f}\frac{\partial f}{\partial t} = -\frac{1}{\mu}$.

Eqs 7.2.6, 7.2.7, 7.2.12, and 7.4.7 provide four equations for the four derivatives $\frac{\partial \Delta p_s}{\partial t}$, $\frac{\partial \Delta p_s}{\partial R}$, $\frac{\partial u_s}{\partial t}$, $\frac{\partial u_s}{\partial R}$ for determining the variation of the shock strength along the shock path (e.g., $\frac{\mathrm{d}\Delta p_s}{\mathrm{d}R}$ where $\frac{\mathrm{d}}{\mathrm{d}R} = \frac{\partial}{\partial R} + \frac{1}{D}\frac{\partial}{\partial t}$).

7.5 The Shock Decay Equation

To obtain the equation for $\frac{\mathrm{d}\Delta p_s}{\mathrm{d}R}$ we first obtain a linear combinations of Eqs 7.2.6, 7.2.7, and 7.4.7 of the form

$$A\frac{\mathrm{d}\Delta p_s}{\mathrm{d}R} + B\frac{\mathrm{d}u_s}{\mathrm{d}R} + C = 0, \tag{7.5.1}$$

where $\frac{\mathrm{d}}{\mathrm{d}R}$ is the total derivative along the shock path. Multiplying Eq 7.2.7 by a multiplier α and Eq 7.4.7 by β and adding the resultant equation with Eq 7.2.6, we obtain after re-arranging the following equation

$$\left(1+\frac{\beta}{u_s}\right)\frac{\partial u_s}{\partial t}+\left(\frac{\partial \rho_s}{\rho_0}\right)\frac{\partial u_s}{\partial R}+\left(\frac{\partial}{\rho_s c_s^2}+\frac{\beta}{\Delta p_s}\right)\frac{\partial \Delta p_s}{\partial t}$$
$$+\frac{1}{\rho_0}\frac{\partial \Delta p_s}{\partial R}+\frac{ju_s}{R}(\alpha+\beta)+\frac{\beta}{\mu}=0.$$

Re-arranging the above equation in a more convenient form to obtain an equation for the variation of Δp_s and u_s along the shock front, we get

$$\frac{\alpha\rho_s}{\rho_0}\left[\frac{\partial u_s}{\partial R}+\frac{\rho_0}{\alpha\rho_s}\left(1+\frac{\beta}{u_s}\right)\frac{\partial u_s}{\partial t}\right]$$
$$+\frac{1}{\rho_0}\left[\left(\frac{\partial \Delta p_s}{\partial R}\right)+\rho_0\left(\frac{\alpha}{\rho_s c_s^2}+\frac{\beta}{\Delta p_s}\right)\frac{\partial \Delta p_s}{\partial t}\right]+(\alpha+\beta)\frac{ju_s}{R}+\beta\mu=0. \tag{7.5.2}$$

For the above equation to be of the form given by Eq 7.5.1, we must choose the multipliers α and β so that

$$\frac{\rho_0}{\alpha\rho_s}\left(1+\frac{\beta}{u_s}\right)=\frac{1}{D}, \tag{7.5.3}$$

$$\rho_0\left(\frac{\alpha}{\rho_s c_s^2}+\frac{\beta}{\Delta p_s}\right)=\frac{1}{D}. \tag{7.5.4}$$

Equation 7.5.2 can then be written in the form

$$\frac{\alpha\rho_s}{\rho_0}\frac{\mathrm{d}u_s}{\mathrm{d}R}+\frac{1}{\rho_0}\frac{\mathrm{d}\Delta p_s}{\mathrm{d}R}+(\alpha+\beta)\frac{ju_s}{R}+\beta\mu=0, \tag{7.5.5}$$

where

$$\frac{\mathrm{d}}{\mathrm{d}R}=\frac{\partial}{\partial R}+\frac{1}{D}\frac{\partial}{\partial t}. \tag{7.5.6}$$

Solving for α and β from Eqs 7.5.3 and 7.5.4 gives

$$\alpha=\frac{\left(D+\frac{\Delta p_s}{u_s\rho_0}\right)}{\left(\frac{D\Delta p_s}{u_s\rho_s c_s^2}+\frac{\rho_s}{\rho_0}\right)}, \tag{7.5.7}$$

$$\beta=\frac{\frac{1}{\rho_s}\left(\frac{\alpha}{c_s^2}-\frac{1}{\alpha}\right)}{\left(\frac{1}{\alpha u_s\rho_s}-\frac{1}{\Delta p_s}\right)}. \tag{7.5.8}$$

Equation 7.5.5 contains derivatives of Δp_s and u_s. Using Eq 7.2.12 to eliminate $\frac{du_s}{dR}$, we obtain

$$\left[D+\frac{\alpha\rho_s}{\rho_0}\left(1-\frac{u_s}{D}\left(\frac{\gamma+1}{4}\right)\right)\right]\frac{\mathrm{d}\Delta p_s}{\mathrm{d}R}+\rho_0 D\left[(\alpha+\beta)\frac{ju_s}{R}+\beta\mu\right]=0.$$

Solving for $\frac{\mathrm{d}\Delta p_s}{\mathrm{d}R}$ yields

$$\frac{\mathrm{d}\Delta p_s}{\mathrm{d}R} = \frac{-\rho_0 D\left[(\alpha+\beta)\frac{ju_s}{R} + \beta\mu\right]}{D + \frac{\alpha\rho_s}{\rho_0}\left(1 - \frac{u_s}{D}\left(\frac{\gamma+1}{4}\right)\right)}. \tag{7.5.9}$$

In the above equation, α and β are given by Eqs 7.5.7 and 7.5.8, and μ is given by Eq 7.4.4 as

$$\mu = \frac{E(R)}{k_j R^j \Delta p_s u_s \nu(R)},$$

$\frac{\rho_s}{\rho_0}$, $\frac{u_s}{D}$ and D are given by the Rankine–Hugoniot equations, that is,

$$\frac{\rho_s}{\rho_0} = \frac{2\gamma + (\gamma+1)\frac{\Delta p_s}{p_0}}{2\gamma + (\gamma-1)\frac{\Delta p_s}{p_0}},$$

$$\frac{u_s}{D} = \frac{2\frac{\Delta p_s}{p_0}}{2 + (\gamma+1)\frac{\Delta p_s}{p_0}}, \tag{7.5.10}$$

$$D^2 = c_0^2\left(1 + \frac{\gamma+1}{2\gamma}\frac{\Delta p_s}{p_0}\right).$$

Equation 7.5.9 contains $\mu(R)$ which is given by Eq 7.4.4. Thus another equation is needed for $E(R)$. From Eq 7.3.7 we obtain

$$\frac{\mathrm{d}z}{\mathrm{d}R} = -k_j R^j \frac{\gamma p_0}{\gamma - 1}\left[\left(1 + \frac{\Delta p_s}{p_0}\right)^{\frac{1}{\gamma}} \frac{\rho_0}{\rho_s} - 1\right]. \tag{7.5.11}$$

The negative sign is introduced to account for the fact that $E(R)$ is a decreasing function of R. Equations 7.5.9 and 7.5.11 can be arranged in a more convenient form as

$$\frac{\mathrm{d}z}{\mathrm{d}R} = -F(z)\left[\frac{zj}{R}Q(z) + \frac{(\gamma+1)z^4 R^j \nu p_0 k_j}{4\gamma^2 E}\right], \tag{7.5.12}$$

where $z = \frac{\Delta p_s}{p_0}$ and

$$F(z) = \frac{4\gamma + 2(\gamma+1)z}{\left(1 + \frac{(\gamma+1)z}{2\gamma}\right)\left(8\gamma + 2z(5\gamma+1) + \frac{z^2(\gamma+1)(5\gamma-1)}{2\gamma}\right)}, \tag{7.5.13}$$

$$Q(z) = 1 + \frac{(3\gamma-1)z}{2\gamma} + \frac{z^2(2\gamma^2 - \gamma + 1)}{4\gamma^2}, \tag{7.5.14}$$

Equation 7.5.12 can be written as

$$\frac{\mathrm{d}E}{\mathrm{d}R} = -\frac{k_j R^j \gamma p_0 G(z)}{(\gamma - 1)}, \tag{7.5.15}$$

where

$$G(z) = (1+z)^{\frac{1}{\gamma}} \left(\frac{2\gamma + (\gamma-1)z}{2\gamma + (\gamma+1)z} - 1 \right). \tag{7.5.16}$$

Given appropriate initial conditions, Eqs 7.5.12 and 7.5.15 can be integrated simultaneously. The energy integral is given by the empirical relationship shown in Eq 7.4.6.

To integrate Eqs 7.5.12 and 7.5.15 one can specify the value of $z(R_i)$ and $E(R_i)$ at some initial radius R_i. Alternately we can also specify $z(R_i)$ and $\left(\frac{\mathrm{d}z}{\mathrm{d}R}\right)_{R_i}$ where $\left(\frac{\mathrm{d}z}{\mathrm{d}R}\right)_{R_i}$ is the initial rate of decay of the shock. If $\frac{\mathrm{d}z}{\mathrm{d}R}$ is specified, E can be determined from Eq 7.5.12. If we specify $\frac{\mathrm{d}z}{\mathrm{d}R}$, then it must be specified to a high degree of accuracy whereas the solution is less sensitive to the energy E.

Note that the energy integral $\nu(z)$ is specified by a relationship that is independent of the solution which in reality should be determined from the solution of the flow field itself. That the BK theory is quite accurate is due to the fact that ν has a small range of values and is a slowly varying function of t. The BK theory is formulated in such a way that the parameter that depends on the flow field can be decoupled from the solution thus permitting the shock decay to be determined without solving for the flow field.

To illustrate the solution of the BK theory, we shall use it to describe the decay of a point blast wave. For a strong spherical blast, the decay is given by the similarity solution descried previously in Chapter 4. As the shock decays to finite shock strengths, the perturbation solution can provide an accurate solution when $\eta \gg 1$ (where $\eta = \frac{1}{M_S^2}$). We shall use the third-order perturbation solution given by Bach and Lee (1969) to provide the initial conditions to start the integration of the BK equations. For convenience we shall write Eqs 7.5.12 and 7.5.15 in non-dimensional form. Defining the explosion length $R_0 = \left(\frac{W_0}{P_0}\right)^{\frac{1}{j+1}}$, where W_0 is the blast energy, we normalize R by R_0, that is $x = \frac{R}{R_0}$. We also express the non-dimensional form of energy W by $\varepsilon = \frac{W}{W_0}$ (note that $E = \gamma W$).

Equations 7.5.12 and 7.5.15 then become

$$\frac{dz}{dx} = -F(z)\left[\frac{zjQ(z)}{x} + \frac{(\gamma+1)z^4 x^j \upsilon k_j}{4\gamma^3}\right], \tag{7.5.17}$$

$$\frac{d\varepsilon}{dx} = -\frac{k_j x^j G(z)}{(\gamma-1)}, \tag{7.5.18}$$

where $F(z)$, $Q(z)$, $G(z)$, and υ are given by Eqs 7.5.13, 7.5.14, 7.5.16, and 7.4.6 respectively.

Using the third-order perturbation solution of Bach and Lee (1969), we compute the initial value of z and $\frac{dz}{dx}$ to start the integration of the BK equations. Three different initial values of $z(x_i)$ and $\frac{dz(x_i)}{dx}$ are chosen corresponding to the shock Mach numbers of $M_{si} = 3.16$, 2.58, and 1.82. The decay of the shock overpressure with distance is shown in Fig. 7.3. The numerical predictions of Goldstine and von Neumann (1955) are also shown in the figure. As can be observed, excellent agreement is obtained.

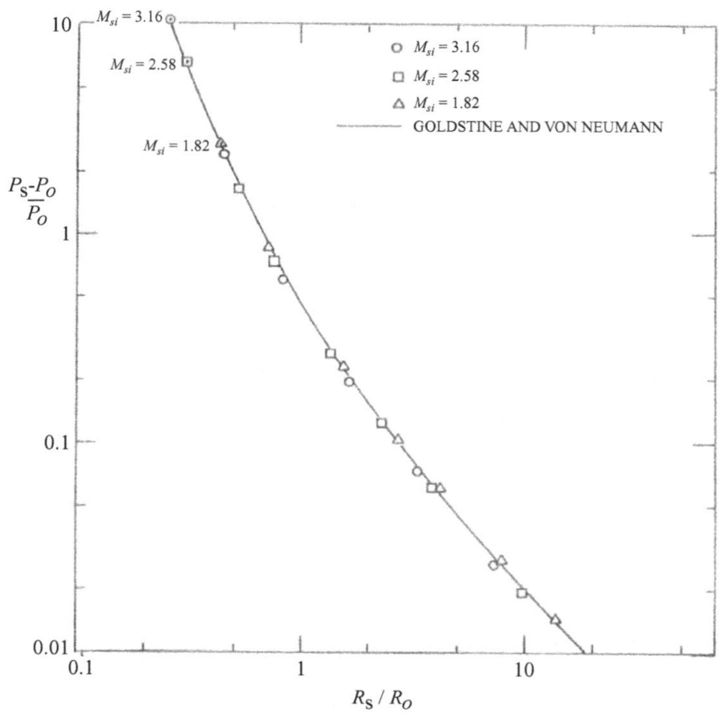

Figure 7.3 Decay of Shock Overpressure with Distance

The BK solutions using different initial conditions are also found to agree with one another.

7.6 The Asymptotic Weak Shock Regime

In the far field where the shock has decayed to a weak shock, that is, $M_s \to 1$, $z \to 0$, Eqs 7.5.12 and 7.5.15 can be simplified permitting them to be integrated. When $z \ll 1$, $\nu \to \frac{2}{3}$, $F(z) \to \frac{1}{2}$, and $Q(z) \to 1$. Equation 7.5.12 simplifies to

$$\frac{\mathrm{d}z}{\mathrm{d}R} = -\frac{jz}{2R} - \frac{(\gamma+1)z^4 R^j \gamma p_0 k_j}{12\gamma^3 E}. \tag{7.6.1}$$

Expanding the function $G(z)$ for small z and retaining the lowest-order term, we get

$$G(z) \simeq \frac{(\gamma-1)(\gamma+1)z^3}{12\gamma^3},$$

and thus Eq 7.5.15 becomes

$$\frac{\mathrm{d}E}{\mathrm{d}R} = -\frac{k_j R^j \gamma p_0 (\gamma+1) z^3}{12\gamma^3}. \tag{7.6.2}$$

Combining Eqs 7.6.1 and 7.6.2, we obtain

$$\frac{\mathrm{d}z}{z} = -\frac{j\mathrm{d}R}{2R} + \frac{\mathrm{d}E}{E},$$

which integrates to yield

$$E = CzR^{\frac{j}{2}}, \tag{7.6.3}$$

where C is a constant of integration. Substituting the solution for E into Eq 7.6.1 gives

$$\frac{\mathrm{d}z}{\mathrm{d}R} = -\frac{jz}{2R} - Kz^3R^{\frac{j}{2}}, \tag{7.6.4}$$

where

$$K = \frac{(\gamma+1)k_j\gamma p_0}{C12\gamma^3}.$$

Equation 7.6.4 can be solved by setting

$$z = \frac{z'}{R^{\frac{j}{2}}},$$

for which it becomes

$$\frac{\mathrm{d}z'}{\mathrm{d}R} = -kz'^3R^{-\frac{j}{2}}, \tag{7.6.5}$$

and for $j \neq 2$, the above equation yields

$$z' = \frac{1}{\left(\frac{4K}{2-j}R^{1-\frac{j}{2}} - C_2\right)^{\frac{1}{2}}},$$

where C_2 is another constant of integration. Thus

$$z = \frac{z'}{R^{\frac{j}{2}}} = \frac{R^{-\frac{j}{2}}}{\left(\frac{4K}{2-j}R^{1-\frac{j}{2}} + C_2\right)^{\frac{1}{2}}}. \tag{7.6.6}$$

For planar geometry where $j = 0$, Eq 7.6.6 gives the asymptotic pressure decay as

$$z = \frac{\Delta p_s}{p_0} \sim \frac{1}{R^{\frac{1}{2}}}. \tag{7.6.7}$$

For the cylindrical geometry where $j = 1$, Eq 7.6.5 gives

$$z \sim \frac{\Delta p_s}{p_0} \sim R^{-\frac{3}{4}}. \tag{7.6.8}$$

For the spherical case ($j = 2$), Eq 7.6.5 becomes

$$\frac{\mathrm{d}z'}{z'^3} = -K\frac{\mathrm{d}R}{R},$$

which integrates to yield

$$\frac{1}{z'^2} = 2K \ln R + C_3,$$

or
$$z = \frac{1}{R(2K \ln R + C_3)^{\frac{1}{2}}},$$

and hence
$$z = \frac{\Delta p_s}{p_0} \sim \frac{1}{R(\ln R)^{\frac{1}{2}}}.$$

The asymptotic decay law from the BK theory is in accord with those described previously in Chapter 2.

7.7 Explosion of a Pressurized Sphere

Consider a sphere of radius R_i with pressure p_i and density ρ_i. The initial strength of the shock formed is z_i when the sphere ruptures and can be found from the non-dimensional shock tube theory (Chapter 5) if the state outside the sphere is p_0, ρ_0, and T_0. Using a perturbation theory, the one-dimensional shock tube flow can be made to account for the spherical geometry and the initial rate of decay of the shock $\frac{dz(R_i)}{dR}$ can be found (McFodden, 1952).

Equation 7.5.12 can be used to find the explosion energy W_0 and hence the explosion length $R_0 = \left(\frac{W_0}{P_0}\right)^{\frac{1}{3}}$ can be obtained. Equations 7.5.17 and 7.5.18 can then be integrated numerically for increasing shock radius $x = \frac{R}{R_0}$. Equivalently, if the explosion energy can be specified directly, the more involved task for using a perturbation solution of the shock tube flow to find $\frac{dz(R_i)}{dR}$ can be avoided. In the BK theory, the explosion energy corresponds to the work done by the expanding interface and requires the solution of the blast wave flow field. However, in the numerical study by Brode (1955), it was found that the internal energy of the sphere given by

$$E_S = \frac{4\pi}{3} R_i^3 \frac{(p_i - p_0)}{\gamma - 1}$$

can be used to represent the blast energy W_0, and the explosion length is $R_0 = \left(\frac{W_0}{p_0}\right)^{\frac{1}{3}}$. With the initial shock strength $z(x_i)$, where $x_i = \frac{R_i}{R_0}$ determined from the shock tube theory, Eqs 7.5.17 and 7.5.18 can be readily integrated using the initial condition $x_i = 1, z(x_i)$, and $\varepsilon(x_i) = 1$. The results for the variation of the shock overpressure $\frac{\Delta p_s}{p_0} = \frac{p_s - p_0}{p_0}$ with shock radius $x = \frac{R}{R_0}$ for various values of the initial pressure ratios $\frac{p_i}{p_0}$ of the sphere are shown in Fig. 7.4. For comparison, the numerical solution for *point* blast decay obtained by Goldstine and von Neumann (1955) is also included. For high values of sphere pressures $(p_s - p_0)/p_0 \gg 1$, the BK solution is found to approach that of the point blast solution. For lower sphere pressures and weaker blasts, the decay is slower and in the far field $x \gg 1$, the shock overpressure is higher than that of the point blast at

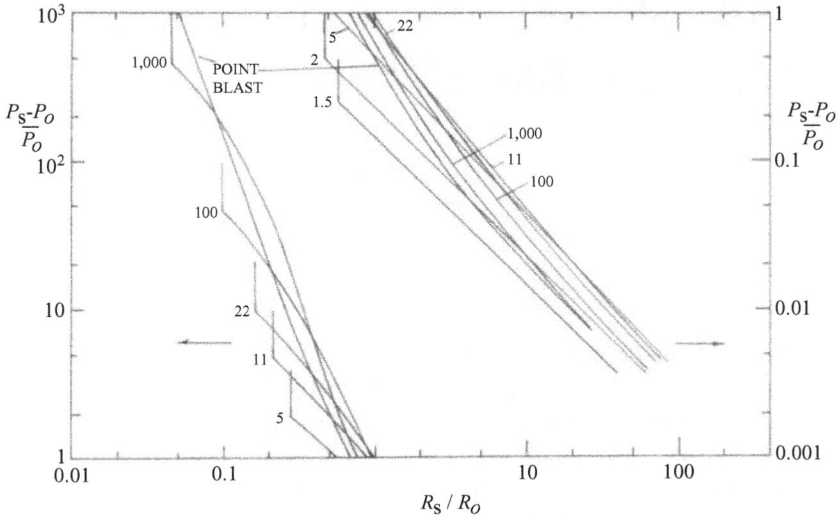

Figure 7.4 Variation of Shock Overpressure with Radius for Different Values of Initial Pressure Ratios

the same shock radius x. This indicates that for strong blast, the energy is dissipated and goes to heating of the core of the blast sphere and less energy is available to drive the shock in the far field for large radius. The BK theory describes the shock front decay and does not provide the flow field behind the blast wave.

8 Non-similar Solutions for Finite Strength Blast Waves

8.1 Introduction

When a blast wave has decayed to finite strength, self-similar solutions are no longer possible. There are a number of analytical methods to describe the propagation of finite strength blast waves and we shall discuss a few of them in this chapter.

If the shock Mach number is not too close to unity and $\frac{1}{M_s^2} \ll 1$, perturbation of the self-similar solution in powers of $\eta = \frac{1}{M_s^2}$ is possible. The zeroth-order solution where $\eta = 0$ would correspond to the self-similar solution for strong blasts. Perturbation solutions in η, which is the inverse of shock Mach number squared, has been obtained by Sakurai (1953, 1954) and others. A quasi-similar or local similar solution was later developed for finite strength blast waves by Oshima (1962). In the quasi-similar method it is assumed that, for a small range of shock strength, the distributions behind the blast are approximately similar. Thus, for a local value of the shock Mach number self-similar solution can be obtained. An integral method has also been developed for finite strength blast waves by Bach and Lee (1970). The integral method is in the same spirit as the Karman–Pohlhausen integral method in boundary-layer theory. By assuming the form for the density profile, the continuity equation can be integrated to yield the particle velocity profile. With the density and velocity profile obtained, integration of the momentum equation then gives the pressure profile. With the three profiles determined, the energy integral can then be used to solve for the shock decay. The three analytical methods mentioned above have their "pros" and "cons" and their choice depends on the application and accuracy desired.

8.2 Basic Formulation

For finite strength blast waves, the variables are functions of both $\xi = \frac{r}{R_s}$ and $\eta = \frac{1}{M_s^2}$ (where ξ and η are used as independent variables instead of r and t). The non-dimensional dependent variables are defined as

$$\psi(\xi, \eta) = \frac{\rho(r, t)}{\rho_0},$$

$$\phi(\xi, \eta) = \frac{u(r, t)}{\dot{R}_s},$$

$$f(\xi, \eta) = \frac{p(r, t)}{\rho_0 \dot{R}_s^2},$$

where $\dot{R}_s = \frac{dR_s}{dt}$ is the shock velocity. The basic conservation equations in the above variables are

$$(\phi - \xi)\frac{\partial \psi}{\partial \xi} + \psi\frac{\partial \phi}{\partial \xi} + \frac{j\phi\psi}{\xi} = 2\theta\eta\frac{\partial \psi}{\partial \eta}, \quad (8.2.1)$$

$$(\phi - \xi)\frac{\partial \phi}{\partial \xi} + \theta\phi + \frac{1}{\psi}\frac{\partial f}{\partial \xi} = 2\theta\eta\frac{\partial \phi}{\partial \eta}, \quad (8.2.2)$$

$$(\phi - \xi)\frac{\partial f}{\partial \xi} + \gamma f\frac{\partial \phi}{\partial \xi} + 2\theta f + \frac{\gamma j\phi f}{\xi} = 2\theta\eta\frac{\partial f}{\partial \eta}, \quad (8.2.3)$$

where

$$\theta(\eta) = \frac{R_s\ddot{R}_s}{\dot{R}_s^2}.$$

The energy integral is written as

$$E_0 = \int_0^{R_s}\left(\frac{\rho u^2}{2} + \frac{p}{\gamma - 1}\right)k_j r^j dr - \int_0^{R_s}\rho_0 e_0 k_j r^j dr,$$

where $e_0 = \frac{p_0}{\rho_0(\gamma-1)}$ is the internal energy of the medium ahead of the shock front. In terms of the blast wave variables ψ, ϕ, f, ξ, η, the energy integral can be written as

$$E_0 = \rho_0\dot{R}_s^2R_s^{j+1}k_jI - \frac{p_0k_jR_s^{j+1}}{(\gamma - 1)(j + 1)}, \quad (8.2.4)$$

where we have used $e_0 = \frac{p_0}{\rho_0(\gamma-1)}$. We define an explosion length R_0 as

$$R_0 = \left(\frac{E_0}{k_j\rho_0c_0^2}\right)^{\frac{1}{j+1}} = \left(\frac{E_0}{k_j\gamma p_0}\right)^{\frac{1}{j+1}}. \quad (8.2.5)$$

Note that the usual definition of explosion length is

$$R_0 = \left(\frac{E_0}{p_0}\right)^{\frac{1}{j+1}}.$$

However, in the definition given by Eq 8.2.5, the constant γk_j is absorbed into the definition of R_0 and this results in a simplification of the energy integral. We shall use the definition of the explosion length given by Eq 8.2.5 here. The energy integral (Eq 8.2.4) now becomes

$$\frac{E_0}{k_j\rho_0c_0^2} = R_0^{j+1} = R_s^{j+1}M_s^2I - \frac{R_s^{j+1}}{\gamma(\gamma - 1)(j + 1)}, \quad (8.2.6)$$

where

$$I = \int_0^1\left(\frac{\psi\phi^2}{2} + \frac{f}{\gamma - 1}\right)\xi^j d\xi.$$

Defining a non-dimensional shock radius y as

$$y = \left(\frac{R_s}{R_0}\right)^{j+1}, \quad (8.2.7)$$

the energy integral can be written as

$$1 = I\frac{y}{\eta} - \frac{y}{\gamma(\gamma-1)(j+1)}. \tag{8.2.8}$$

When the shock is strong, $M_s \gg 1$, $\eta \ll 1$, the first term on the right hand side dominates; thus

$$1 \simeq I\frac{y}{\eta} \quad \text{for} \quad \eta \ll 1$$

Since the left hand side of the above equation is unity, that is, $\frac{y}{\eta}$ must approach a constant as $\eta \to 0$. Hence $y \to A\eta$ where A is a constant. From the definition of y and η, we see that $\frac{y}{\eta} \to A$ means

$$\left(\frac{R_s}{R_0}\right)^{j+1}\frac{\dot{R}_s^2}{c_0^2} \to A,$$

and with $R_s \sim t^N$, $R_s^{j+1}\dot{R}_s^2 \sim t^{(j+1)N+2(N-1)}$ and the exponent of time t must vanish since the right hand side is a constant. This gives $N = \frac{2}{j+3}$ and $\theta = -\left(\frac{j+1}{2}\right)$ and the self-similar solution for strong blast is recovered.

The energy integral given by Eq 8.2.8 introduces a new variable y (Eq 8.2.7), thus another equation is required for y. From the definition of y, we can differentiate and obtain

$$\frac{dy}{dR_s} = \frac{(j+1)}{R_0}\left(\frac{R_s}{R_0}\right)^j,$$

and since $R_s\frac{\partial}{\partial R_s} = -2\theta\eta\frac{\partial}{\partial\eta}$, the above equation becomes

$$\frac{dy}{d\eta} = -\left(\frac{j+1}{2}\right)\left(\frac{y}{\theta\eta}\right). \tag{8.2.9}$$

Equations 8.2.1, 8.2.2, 8.2.3, 8.2.8, and 8.2.9 constitute five equations for the five dependent variables, ψ, ϕ, f, η, and y.

The boundary conditions at the shock front $\xi = 1$ are given by the Rankine–Hugoniot equations as

$$\psi(1,\eta) = \frac{\gamma+1}{(\gamma-1)+2\eta}, \tag{8.2.10}$$

$$\phi(1,\eta) = \frac{2}{\gamma+1}(1-\eta), \tag{8.2.11}$$

$$f(1,\eta) = \frac{2}{\gamma+1}(1-\frac{\gamma-1}{2\gamma}\eta). \tag{8.2.12}$$

The solution $\psi(\xi,\eta)$, $\phi(\xi,\eta)$, etc. must also satisfy the rear boundary condition of $\phi = 0$ at the center of symmetry $\xi = 0$. The initial conditions are at $t = 0$, $y = 0$, $\eta = 0$.

Equations 8.2.1–8.2.3 are still partial differential equations like the Euler's equations. The blast wave variables replace the ρ, u, p, r, t of the dimensional Euler's equations.

8.3 Perturbation Solution

For strong and moderate strength shock waves where $\eta \ll 1$, we may seek a perturbation solution in the form of a power series in η (Sakurai 1953, 1954; Swigart 1960; Korobenikov and Clushkin 1963; Bach and Lee 1969). Accordingly we write

$$\left.\begin{aligned} \phi(\xi,\eta) &= \Sigma\phi_n\eta^n = \phi_0(\xi) + \phi_1(\xi)\eta + \phi_2(\xi)\eta^2 + \cdots \\ \psi(\xi,\eta) &= \Sigma\psi_n\eta^n = \psi_0(\xi) + \psi_1(\xi)\eta + \psi_2(\xi)\eta^2 + \cdots \\ f(\xi,\eta) &= \Sigma f_n\eta^n = f_0(\xi) + f_1(\xi)\eta + f_2(\xi)\eta^2 + \cdots \end{aligned}\right\} \tag{8.3.1}$$

where the zeroth-order solution is the self-similar constant energy solution. The shock radius y can also be expressed as a power series in η as

$$y(\eta) = \Sigma A_n\eta^n = A_1\eta + A_2\eta^2 + A_3\eta^3 + \cdots \tag{8.3.2}$$

The form above follows from the energy integral where $\frac{y}{\eta} \to$ constant as $\eta \to 0$. Similarly, the energy integral can be expanded using Eq 8.3.1, that is,

$$I = \int_0^1 \left(\frac{\psi\phi^2}{2} + \frac{f}{\gamma-1}\right)\xi^j \mathrm{d}\xi = I_0 + I_1\eta + I_2\eta^2 + \cdots$$

where

$$\begin{aligned} I_0 &= \int_0^1 \left(\frac{\psi_0\phi_0^2}{2} + \frac{f_0}{\gamma-1}\right)\xi^j \mathrm{d}\xi, \\ I_1 &= \int_0^1 \left(\frac{\psi_1\phi_0^2}{2} + \psi_0\phi_0\phi_1 + \frac{f_1}{\gamma-1}\right)\xi^j \mathrm{d}\xi, \\ I_2 &= \int_0^1 \left(\frac{\psi_2\phi_0^2}{2} + \frac{\psi_0\phi_1^2}{2} + \psi_1\phi_1\phi_0 + \psi_0\phi_0\phi_2 + \frac{f_2}{\gamma-1}\right)\xi^j \mathrm{d}\xi, \\ I_3 &= \int_0^1 \left(\frac{\psi_3\phi_0^2}{2} + \frac{\psi_1\phi_1^2}{2} + \psi_0\phi_0\phi_2 + \psi_0\phi_1\phi_2 + \psi_1\phi_0\phi_2 \right. \\ &\quad \left. + \psi_2\phi_0\phi_1 + \frac{f_3}{\gamma-1}\right)\xi^j \mathrm{d}\xi. \end{aligned} \tag{8.3.3}$$

The shock decay coefficient, $\theta(\eta)$ can be also expanded as

$$\theta(\eta) = \Sigma\theta_n\eta^n = \theta_0 + \theta_1\eta + \theta_2\eta^2 + \theta_3\eta^3 + \cdots$$

where

$$\begin{aligned} \theta_0 &= -\left(\frac{j+1}{2}\right), \\ \theta_1 &= \frac{j+1}{2}\frac{A_2}{A_1}, \\ \theta_2 &= (j+1)\left(\frac{A_3}{A_1} - \left(\frac{A_2}{A_1}\right)^2\right), \\ \theta_3 &= \frac{j+1}{2}\left(\frac{4A_2^3}{A_1^3} - \frac{7A_2A_3}{A_1^2} + \frac{3A_4}{A_1}\right). \end{aligned} \tag{8.3.4}$$

The above expansions for $\theta(\eta)$ are obtained from Eq 8.2.9 using the expansion for $y(\eta)$ given by Eq 8.3.2. Since the energy integral relates y with I, the coefficients of the expansion for $y(\eta)$ in Eq 8.3.2 can be expressed in terms of the energy integral, that is, I_0, I_1, I_2, etc. as

$$\begin{aligned}
A_1 &= \frac{1}{I_0},\\
A_2 &= \frac{1}{I_0}\left[\frac{A_1}{\gamma(\gamma-1)(j+1)} - A_1 I_1\right],\\
A_3 &= \frac{1}{I_0}\left[\frac{A_2}{\gamma(\gamma-1)(j+1)} - A_2 I_1 - I_2 A_1\right],\\
A_4 &= \frac{1}{I_0}\left[\frac{A_3}{\gamma(\gamma-1)(j+1)} - A_1 I_3 - A_2 I_2 - A_3 I_1\right].
\end{aligned} \tag{8.3.5}$$

The equations for the different order solutions can be obtained by substituting the expansions for ψ, ϕ, f, θ into the basic equations (i.e., Eqs 8.2.1–8.2.3). Collecting terms of the same order of magnitude in η, we obtain for

<u>Zeroth order in η</u>

$$(\phi_0 - \xi)\psi_0' + \psi_0\phi_0' = -\frac{j\phi_0\psi_0}{\xi} = \Lambda_1^{(0)}, \tag{8.3.6}$$

$$(\phi_0 - \xi)\phi_0' + \frac{f_0'}{\psi_0} = -\theta_0\phi_0 = \Lambda_2^{(0)}, \tag{8.3.7}$$

$$(\phi_0 - \xi)f_0' + \gamma f_0\phi_0' = -\left(2\theta_0 f_0 + \frac{\gamma j f_0\phi_0}{\xi}\right) = \Lambda_3^{(0)}. \tag{8.3.8}$$

<u>First order in η</u>

$$\begin{aligned}
&(\phi_0 - \xi)\psi_1' + \psi_0\phi_1' = \\
&-\left[\psi_1(\phi_0' - 2\theta_0) + \phi_1\psi_0' + \frac{j(\psi_0\phi_1 + \psi_1\phi_0)}{\xi}\right] = \Lambda_1^{(1)},
\end{aligned} \tag{8.3.9}$$

$$\begin{aligned}
&(\phi_0 - \xi)\phi_1' + \frac{f_1'}{\psi_0} = \\
&-\left[\phi_1(\phi_0' - \theta) + \theta_1\phi_0 - \frac{\psi_1 f_0'}{\psi_0^2}\right] = \Lambda_2^{(1)},
\end{aligned} \tag{8.3.10}$$

$$\begin{aligned}
&(\phi_0 - \xi)f_1' + \gamma f_0\phi_1' = \\
&-\left[\gamma f_1\phi_0' + \phi_1 f_0' + 2\theta_1 f_0 + \frac{\gamma j(f_0\phi_1 + f_1\phi_0)}{\xi}\right] = \Lambda_3^{(1)}.
\end{aligned} \tag{8.3.11}$$

<u>Second order in η</u>

$$
\begin{aligned}
(\phi_0 - \xi)\psi_2' + \psi_0\phi_2' = &- \Bigg[\psi_2(\phi_0' - 4\theta_0) + \psi_1\phi_1' + \phi_1\psi_1' + \phi_2\psi_0' - 2\psi_1\theta_1 \\
&+ \frac{j(\psi_0\theta_2 + \psi_1\phi_1 + \psi_2\phi_0)}{\xi}\Bigg] = \Lambda_1^{(2)},
\end{aligned} \tag{8.3.12}
$$

$$
\begin{aligned}
(\phi_0 - \xi)\phi_2' + \frac{f_2'}{\psi_0} = &- \Bigg[\phi_2(\phi_0' - 3\theta_0) + \phi_1\phi_1' - \theta_1\phi_1 + \theta_2\phi_0 \\
&- \frac{\psi_1 f_1'}{\psi_0^2} + \frac{f_0'}{\psi_0^2}\left(\frac{\psi_1^2}{\psi_0} - \psi_2\right)\Bigg] = \Lambda_2^{(2)},
\end{aligned} \tag{8.3.13}
$$

$$
\begin{aligned}
(\phi_0 - \xi)f_2' + \gamma f_0\phi_2' = &- \Bigg[f_2(\gamma\phi_0' - 2\theta_0) + \gamma f_1\phi_1' + \phi_1 f_1' + \phi_2 f_0' + 2\theta_2 f_0 \\
&+ \frac{\gamma j(f_0\phi_0 + f_1\phi_1 + f_2\phi_0)}{\xi}\Bigg] = \Lambda_3^{(2)},
\end{aligned} \tag{8.3.14}
$$

<u>Third order in η</u>

$$
\begin{aligned}
(\phi_0 - \xi)\psi_3' + \psi_0\phi_3' = &- \Big[\psi_3(\phi_0' - 6\theta_0) + \psi_2(\phi_1' - 4\theta_1) + \psi_1(\phi_2' - 2\theta_2) \\
&+ \phi_3\psi_0' + \phi_2\psi_1' + \phi_1\psi_2' \\
&+ \frac{j}{\xi}(\psi_0\phi_3 + \psi_1\phi_2 + \psi_2\phi_1 + \phi_0\psi_3)\Big] = \Lambda_1^{(3)},
\end{aligned} \tag{8.3.15}
$$

$$
\begin{aligned}
(\phi_0 - \xi)\phi_3' + \frac{f_3'}{\psi_0} = &- \Bigg[\phi_3(\phi_0' - 5\theta_0) + \phi_2(\phi_1' - 3\theta_1) + \phi_1(\phi_2' - \theta_2) \\
&+ \theta_3\phi_0' + \frac{\psi_1 f_2'}{\psi_0^2} + \frac{f_1'}{\psi_0^2}\left(\frac{\psi_1^2}{\psi_0} - \psi_2\right) \\
&- \frac{f_0'}{\psi_0^2}\left(\frac{\psi_3}{\psi_2^2} - \frac{2\psi_1\psi_2}{\psi_0} + \psi_3\right)\Bigg] = \Lambda_2^{(3)},
\end{aligned} \tag{8.3.16}
$$

$$
\begin{aligned}
(\phi_0 - \xi)f_3' + \gamma f_0\phi_3' = &- \Big[f_3(\gamma\phi_0' - 4\theta_0) + f_2(\gamma\phi_1' - 2\theta_1) \\
&+ \gamma f_1\phi_2' + 2\theta_3 f_0 + \phi_3 f_0' + \phi_2 f_1' + \phi_1 f_2' \\
&+ \frac{\gamma j}{\xi}(f_0\phi_3 + f_1\phi_2 + f_2\phi_1 + f_3\phi_0)\Big] = \Lambda_3^{(3)}.
\end{aligned} \tag{8.3.17}
$$

The zeroth-order equations are the self-similar equations for the constant energy strong blast. With $\theta_0 = -\left(\frac{j+1}{2}\right)$, Eqs 8.3.6–8.3.8 can be integrated immediately using the strong shock boundary conditions at the front $\xi = 1$. The integral I_0 (Eq 8.3.5) can then be computed and A_1 determined.

The boundary conditions for the higher-order solutions at the shock can be obtained by expanding the Rankine–Hugoniot equations, that is, Eqs 8.2.10–8.2.12. Hence

$$\phi(1,\eta) = \phi_0(1) + \phi_1(1)\eta + \phi_2(1)\eta^2 + \cdots \tag{8.3.18}$$

where

$$\phi_0(1) = \frac{2}{\gamma+1},\ \phi_1(1) = -\frac{2}{\gamma+1},\ \phi_2(1) = 0,\ \phi_3(1) = 0.$$

For the pressure

$$f(1,\eta) = f_0(1) + f_1(1)\eta + f_2(1)\eta^2 + \cdots \tag{8.3.19}$$

where

$$f_0(1) = \frac{2}{\gamma+1},\ f_1(1) = -\frac{2}{\gamma(\gamma+1)},\ f_2(1) = 0,\ f_3(1) = 0,$$

and similarly for the density

$$\psi(1,\eta) = \psi_0(1) + \psi_1(1)\eta + \psi_2(1)\eta^2 + \cdots \tag{8.3.20}$$

where

$$\psi_0(1) = \frac{\gamma+1}{\gamma-1},\ \psi_1(1) = -\frac{2(\gamma+1)}{(\gamma-1)^2},\ \psi_2(1) = \frac{4(\gamma+1)}{(\gamma-1)^3},\ \psi_3(1) = \frac{\gamma(\gamma+1)}{(\gamma-1)^4}.$$

To facilitate the numerical integration of Eqs 8.3.6–8.3.17, we can solve for the derivatives ψ_0', ψ_1', etc. and obtain

$$\psi_0' = \frac{(\phi_0-\xi)^2\Lambda_1^{(0)} + \Lambda_3^{(0)} - \frac{\gamma f_0}{\psi_0}\Lambda_1^{(0)} - (\phi_0-\xi)\psi_0\Lambda_2^{(0)}}{(\phi_0-\xi)\left[(\phi_0-\xi)^2 - \frac{\gamma f_0}{\psi_0}\right]}, \tag{8.3.21}$$

$$\phi_0' = \frac{(\phi_0-\xi)\Lambda_2^{(0)} - \frac{\Lambda_3^{(0)}}{\psi_0}}{(\phi_0-\xi)^2 - \frac{\gamma f_0}{\psi_0}}, \tag{8.3.22}$$

$$f_0' = \frac{(\phi_0-\xi)\Lambda_3^{(0)} - \gamma f_0\Lambda_2^{(0)}}{(\phi_0-\xi)^2 - \frac{\gamma f_0}{\psi_0}}. \tag{8.3.23}$$

For the first-order derivatives ψ_1', ϕ_1', f_1', we obtain

$$\psi_1' = \frac{(\phi_0-\xi)^2\Lambda_1^{(1)} + \Lambda_3^{(1)} - \frac{\gamma f_0}{\psi_0}\Lambda_1^{(1)} - (\phi_0-\xi)\psi_0\Lambda_2^{(1)}}{(\phi_0-\xi)\left[(\phi_0-\xi)^2 - \frac{\gamma f_0}{\psi_0}\right]}, \tag{8.3.24}$$

$$\phi_1' = \frac{(\phi_0-\xi)\Lambda_2^{(1)} - \frac{\Lambda_3^{(1)}}{\psi_0}}{\left[(\phi_0-\xi)^2 - \frac{\gamma f_0}{\psi_0}\right]}, \tag{8.3.25}$$

$$f_1' = \frac{(\phi_0-\xi)\Lambda_3^{(1)} - \gamma f_0\Lambda_2^{(1)}}{\left[(\phi_0-\xi)^2 - \frac{\gamma f_0}{\psi_0}\right]}. \tag{8.3.26}$$

For the second-order derivatives, we have

$$\psi_2' = \frac{(\phi_0 - \xi)^2 \Lambda_1^{(2)} + \Lambda_3^{(2)} - \frac{\gamma f_0}{\psi_0}\Lambda_1^{(2)} - (\phi_0 - \xi)\psi_0 \Lambda_2^{(2)}}{(\phi_0 - \xi)\left[(\phi_0 - \xi)^2 - \frac{\gamma f_0}{\psi_0}\right]}, \tag{8.3.27}$$

$$\phi_2' = \frac{(\phi_0 - \xi)\Lambda_2^{(2)} - \frac{\Lambda_3^{(2)}}{\psi_0}}{(\phi_0 - \xi)^2 - \frac{\gamma f_0}{\psi_0}}, \tag{8.3.28}$$

$$f_2' = \frac{(\phi_0 - \xi)\Lambda_3^{(2)} - \gamma f_0 \Lambda_2^{(2)}}{(\phi_0 - \xi)^2 - \frac{\gamma f_0}{\psi_0}}. \tag{8.3.29}$$

For the third-order derivatives, we obtain

$$\psi_3' = \frac{(\phi_0 - \xi)^2 \Lambda_1^{(3)} + \Lambda_3^{(3)} - \frac{\gamma f_0}{\psi_0}\Lambda_1^{(3)} - (\phi_0 - \xi)\psi_0 \Lambda_2^{(3)}}{(\phi_0 - \xi)\left[(\phi_0 - \xi)^2 - \frac{\gamma f_0}{\psi_0}\right]}, \tag{8.3.30}$$

$$\phi_3' = \frac{(\phi_0 - \xi)\Lambda_2^{(3)} - \frac{\Lambda_3^{(3)}}{\psi_0}}{(\phi_0 - \xi)^2 - \frac{\gamma f_0}{\psi_0}}, \tag{8.3.31}$$

$$f_3' = \frac{(\phi_0 - \xi)\Lambda_3^{(3)} - \gamma f_0 \Lambda_2^{(3)}}{(\phi_0 - \xi)^2 - \frac{\gamma f_0}{\psi_0}}. \tag{8.3.32}$$

The variables $\Lambda_1^{(n)}$, $\Lambda_2^{(n)}$, and $\Lambda_3^{(n)}$, where $n = 0, 1, 2, 3, \ldots$ are the left hand terms of Eqs 8.3.6–8.3.17 for the various orders in η.

For the higher-order solution, θ_n is initially not known and hence the higher-order equations cannot be integrated as the zeroth-order equations. An iteration procedure is required to determine the higher-order solutions. For example, to determine the first-order solution, a trial value of A_2 is first assumed and θ_1 can then be found from Eq 8.3.4, that is,

$$\theta_1 = \frac{j+1}{2}\frac{A_2}{A_1}.$$

The first-order equations (i.e., Eqs 8.3.24–8.3.26) can now be integrated numerically. The integral I_1 can then be determined and from Eq 8.3.5, a new value of A_2 is found which can be compared to the assumed value of A_2. The correct solution will be obtained when the assumed value of A_2 agree with the computed value from the energy integral to within a prescribed accuracy. In general, it is found that the energy integral is not very sensitive to the solution and hence is used unless a very high degree of accuracy is required. On the other hand the rear boundary condition at the center of symmetry which requires the particle velocity to vanish, is found to provide a much more sensitive criterion for iterating for the correct solution. With an assumed value of A_2, and hence θ_1, the first-order equations are integrated and $\phi_1(0) = 0$ is used as the criterion to arrive at the correct solution. In the study by Bach and Lee (1969), an accuracy of $|\phi_n(0)| < 10^{-9}$ was found to give very good results for the higher-order solutions.

To obtain the shock trajectory, the identity

$$R_s = \int \dot{R}_s \mathrm{d}t,$$

can be used. In terms of y and η, the above equation can be written as

$$\mathrm{d}\left(\frac{c_0 t}{R_0}\right) = \frac{\eta^{\frac{1}{2}} \mathrm{d}y}{(j+1) y^{\frac{j}{j+1}} (\gamma k_j)^{\frac{1}{j+1}}},$$

and substituting the perturbation expressions for y into the above equation and integrating yields

$$\frac{c_0 t}{R_0} = B_1 \eta^{\frac{j+3}{2(j+1)}} (1 + B_2 \eta + B_3 \eta^2 + B_4 \eta^3 + \cdots), \tag{8.3.33}$$

where

$$B_1 = \frac{2 A_1^{\frac{1}{j+1}}}{j+3}, \tag{8.3.34}$$

$$B_2 = \frac{(j+2)(j+3) A_2}{(j+1)(3j+5) A_1}, \tag{8.3.35}$$

$$B_3 = \frac{(2j+3)(j+3)\left[\frac{A_3}{A_1} - \frac{j A_2^2}{2(j+1) A_1^2}\right]}{(j+1)(5j+7)}, \tag{8.3.36}$$

$$B_4 = \frac{(3j+4)(j+3)}{(j+1)(7j+9)}\left[\frac{A_4}{A_1} - \frac{j A_2 A_3}{(j+1) A_1^2} + \frac{j(2j+1) A_2^3}{6(j+1)^2 A_1^3}\right]. \tag{8.3.37}$$

For $\gamma = 1.4$, the perturbation coefficients are given in Table 8.1 (Bach and Lee, 1969) and compared to previous Authors.

The analysis in this section is for a perturbation solution in η. Alternatively one can seek a perturbation solution in the shock radius y, that is, $\psi(\xi, \eta) = \Sigma \psi_r(\xi) y^n = \psi_0(\xi) + \psi_1(\xi) y + \cdots$, etc., since $\eta \ll 1$ corresponds to $y \ll 1$ also. Note that, to the same order in the perturbation parameter, the accuracy of the solution may be different.

Higher-order solutions require the solutions for the lower orders. Thus, a very high accuracy is required for the lower-order solutions. Higher-order terms in the perturbation equation diverges and it is not worthwhile to continue to orders higher than third order.

A comparison between the perturbation solution and finite difference numerical integration of the conservation equations is illustrated in Figs. 8.1–8.3 for spherical blast with $\gamma = 1.4$. In Fig. 8.1, the shock decay coefficient θ is given. The variation of shock strength with shock radius is shown in Fig. 8.2. The shock trajectory is shown in Fig. 8.3.

8.4 Quasi-similar Solution

Oshima (1962) developed a method to describe the decay of finite strength blast waves. He called the solution "quasi-similar" solution. For strong blast waves where $\eta \to 0$,

Table 8.1 Perturbation Coefficients

j	0	1	2
θ_0	−0.50000	−1.00000	−1.50000
θ_1	1.07167	1.98356	2.87723
	1.0717[b]	1.9836[b]	2.8773[b]
	1.068[a]	1.989[a]	2.877[a]
θ_2	−1.36056	−2.10160	−2.84969
		−2.7373[c]	
θ_3	2.08025	2.97464	3.89983
A_1	0.82497	1.59622	2.36246
A_2	1.76819	3.16620	4.53155
A_3	2.66741	4.60305	6.44813
A_4	3.65364	6.27709	8.67653
I_0	1.21217	0.62648	0.42329
I_1	−0.81239	−0.34981	−0.21669
I_2	1.64924	0.65831	0.40208
I_3	−0.50279	−0.18592	−0.10975
B_1	0.54998	0.63171	0.53274
B_2	2.57201	1.48767	1.16252
B_3	4.15718	1.58341	1.03146
B_4	5.90511	1.79201	1.07922

[a] Sakurai (1965).
[b] Korobeinikov and Clushkin (1963).
[c] Swigart (1960).

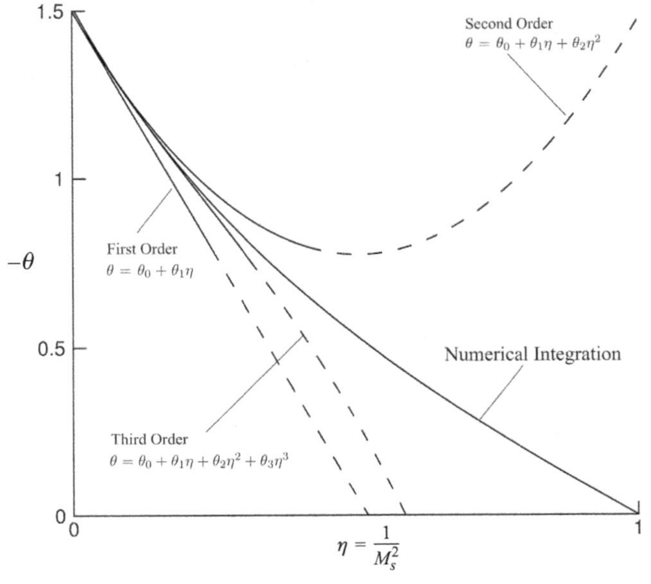

Figure 8.1 Shock Decay Coefficient θ

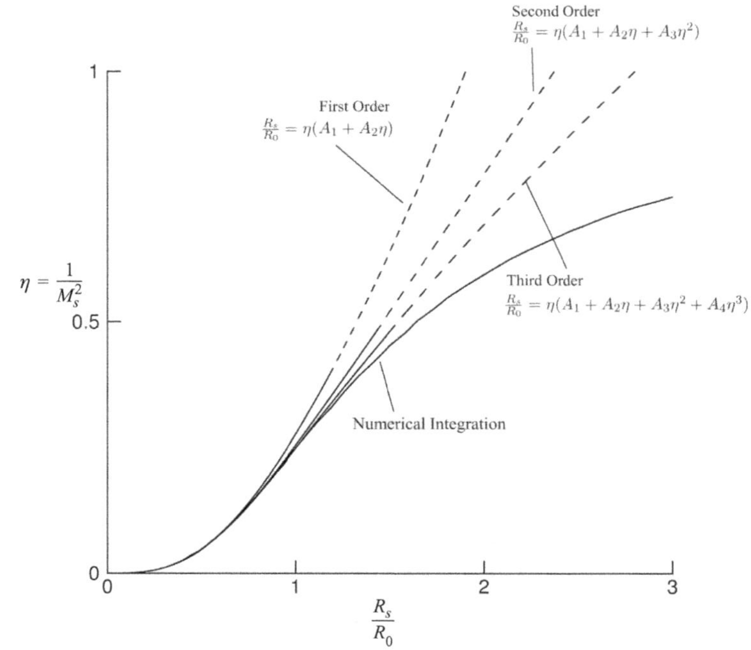

Figure 8.2 The Variation of Shock Strength with Shock Radius

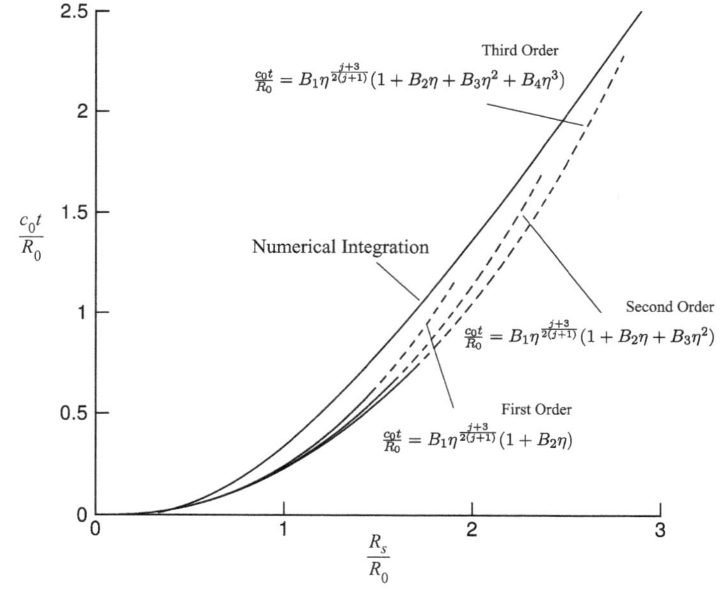

Figure 8.3 The Shock Trajectory

the right hand side of Eqs 8.2.1–8.2.3 vanishes and if θ is not an explicit function of η, then Eqs 8.2.1–8.2.3 reduce to ordinary differential equations giving the self-similar solution $\psi(\xi)$, $\phi(\xi)$, $f(\xi)$ for strong blasts. In the quasi-similar solution of Oshima, the derivatives of ψ, ϕ, f with respect to η on the right hand side of Eqs 8.2.1–8.2.3 are replaced by functions of η so that for a local value of η, Eqs 8.2.1–8.2.3 become ordinary differential equations, that is, the solution is similar for a local value of η. Oshima justified this approximation by noting that the density distribution for nearly equal shock strengths exhibits similar features. The quasi-similar approximation of Oshima for blast wave decay is analogous to the local similarity approximation used for boundary layer over blunt-nosed bodies in hypersonic flow where the derivatives with respect to the streamwise variable are assumed to be small compared to the derivatives normal to the streamline. This results in terms depending only on the streamwise variable arising from the external flow or wall conditions can be assumed to take on their local values. The local-similar approximation represents a patching together of local solutions. The validity of this approximation requires that the external flow properties vary slowly with the streamwise variable. In the blast wave problem, the local similar approximation implies that the variations of ψ, ϕ, f with respect to ξ are much larger than the variation with respect to η. Oshima quoted experimental observations to justify the local similarity approximation.

To render the right hand derivatives of terms in Eqs 8.2.1–8.2.3 with respect to η to be a function of η, Oshima assumed that, for a narrow range of the shock strength, the solutions can be approximated by the form $\psi(\xi,\eta) = \psi_1(\xi)\psi_2(\eta)$, $\phi(\xi,\eta) = \phi_1(\xi)\phi_2(\eta)$, and $f(\xi,\eta) = f_1(\xi)f_2(\eta)$. Differentiating the assumed form of the quasi-similar solution, we obtain

$$\left.\frac{\partial \ln \psi}{\partial \eta}\right)_\xi = \frac{\partial \ln \psi_2(\eta)}{\partial \eta},$$

$$\left.\frac{\partial \ln \phi}{\partial \eta}\right)_\xi = \frac{\partial \ln \phi_2(\eta)}{\partial \eta},$$

$$\left.\frac{\partial \ln f}{\partial \eta}\right)_\xi = \frac{\partial \ln f_2(\eta)}{\partial \eta}.$$

The derivatives with respect to η on the right hand side of the above equation is independent on ξ. Thus we may evaluate them at the shock front where $\xi = 1$ and using the Rankine–Hugoniot equations (i.e., Eqs 8.2.10–8.2.12) we obtain

$$\left.\frac{\partial \ln \psi}{\partial \eta}\right)_\xi = \left.\frac{\partial \ln \psi}{\partial \eta}\right)_{\xi=1} = \frac{\partial \ln \psi(1)}{\partial \eta} = \frac{-2\eta}{(\gamma-1)+2\eta},$$

$$\left.\frac{\partial \ln \phi}{\partial \eta}\right)_\xi = \left.\frac{\partial \ln \phi}{\partial \eta}\right)_{\xi=1} = \frac{\partial \ln \phi(1)}{\partial \eta} = \frac{-\eta}{1-\eta},$$

$$\left.\frac{\partial \ln f}{\partial \eta}\right)_\xi = \left.\frac{\partial \ln f}{\partial \eta}\right)_{\xi=1} = \frac{\partial \ln f(1)}{\partial \eta} = \frac{-(\gamma-1)\eta}{2\gamma-(\gamma-1)\eta}.$$

Replacing the derivatives with respect to η on the right hand side of Eqs 8.2.1–8.2.3 by the above and rearranging yield

$$(\phi - \xi)\psi' + \psi\phi' + \frac{j\phi\psi}{\xi} = \Lambda_1 \psi, \tag{8.4.1}$$

$$(\phi - \xi)\phi' + \frac{f'}{\psi} = \Lambda_2 \phi, \tag{8.4.2}$$

$$(\phi - \xi)f' + \gamma f\phi' + \frac{\gamma j f\phi}{\xi} = \Lambda_3 f, \tag{8.4.3}$$

where

$$\Lambda_1 = \frac{-4\theta\eta}{(\gamma - 1) + 2\eta}, \tag{8.4.4}$$

$$\Lambda_2 = -\theta\left(\frac{2\eta}{1 - \eta} + 1\right), \tag{8.4.5}$$

$$\Lambda_3 = -2\theta\left(\frac{(\gamma - 1)\eta}{2\gamma - (\gamma - 1)\eta} + 1\right). \tag{8.4.6}$$

For a local value of the shock strength η, Eqs 8.4.1–8.4.3 then become a set of ordinary differential equations which can be integrated using the boundary condition at the shock front $\xi = 1$ given by Eqs 8.2.10–8.2.12.

To facilitate the integration, we can solve for the derivatives ψ', ϕ', and f' from Eqs 8.4.1–8.4.3 and obtain

$$\psi' = \frac{(\phi - \xi)(-\frac{j\phi\psi}{\xi} + \Lambda_1\psi) + \Lambda_3 f - \frac{\gamma j\phi f}{\xi} - \frac{\gamma f}{\psi}(\frac{-j\phi\psi}{\xi} + \Lambda_1\psi) - \frac{(\phi-\xi)\Lambda_2\phi}{\psi}}{(\phi - \xi)((\phi - \psi)^2 - \frac{\gamma f}{\psi})}, \tag{8.4.7}$$

$$\phi' = \frac{(\phi - \xi)\Lambda_2\phi - (\Lambda_3 f - \frac{\gamma j f\phi}{\xi})\frac{1}{\psi}}{(\phi - \xi)^2 - \frac{\gamma f}{\psi}}, \tag{8.4.8}$$

$$f' = \frac{\gamma f\phi\Lambda_2 - (\phi - \xi)(\Lambda_3 f - \frac{\gamma j f\phi}{\xi})}{(\phi - \xi)^2 - \frac{\gamma f}{\psi}}. \tag{8.4.9}$$

Note that as in the case of the similarity equations for strong blast waves, a first integral can also be obtained from the quasi-similarity equations (i.e., Eqs 8.4.1–8.4.3). Multiplying Eq 8.4.1 by $[(\phi - \xi)\psi]^{-1}$ and Eq 8.4.3 by $[(\phi - \xi)f]^{-1}$ and arranging the results:

$$\frac{\psi'}{\psi} + \frac{(\phi - \xi)'}{(\phi - \xi)} + \frac{j + 1 + \Lambda_1}{(\phi - \xi)} + \frac{j}{\xi} = 0,$$

$$\frac{f'}{f} + \frac{\gamma(\phi - \xi)'}{(\phi - \xi)} + \frac{\gamma(j + 1) + \Lambda_3}{(\phi - \xi)} + \frac{\gamma j}{\xi} = 0.$$

Eliminating $(\phi - \xi)$ among the above expression yields

$$\chi_2 \frac{f'}{f} - \chi_1 \frac{\psi'}{\psi} + (\gamma\chi_2 - \chi_1)\frac{(\phi-\xi)'}{(\phi-\xi)} + \frac{\gamma j \chi_2 - j\chi_1}{\xi} = 0,$$

which integrates to give the first integral as

$$\frac{f^{\chi_2}(\phi-\xi)^{\gamma\chi_2-\chi_1}\xi^{\gamma\chi_2-\chi_1}}{\psi^{\chi_1}} = C(\eta), \tag{8.4.10}$$

where

$$\chi_1 = \gamma(j+1) - \Lambda_3,$$
$$\chi_2 = (j+1) - \Lambda_1.$$

The constant of integration $C(\eta)$ can be evaluated at the shock front $\xi = 1$ using the Rankine–Hugoniot relations (Eqs 8.2.10–8.2.12). The first integral given by Eq 8.4.10 can be expressed in a more convenient form as

$$\psi^{\gamma}[\psi(\xi-\phi)\xi^{j}]^{\delta} = C_1(\eta)f, \tag{8.4.11}$$

where $\delta = \frac{-(\gamma\chi_2-\chi_1)}{\chi_2} = \frac{\gamma\Lambda_1-\Lambda_3}{j+1-\Lambda_1}$.
Substituting Λ_1, Λ_3 from Eqs 8.4.4 and 8.4.6 we obtain

$$\delta = \frac{4\theta\gamma\left(\frac{1}{2\gamma-(\gamma-1)\eta} - \frac{\eta}{(\gamma-1)+2\eta}\right)}{(j+1) + \frac{4\theta\eta}{(\gamma-1)+2\eta}}. \tag{8.4.12}$$

Using the Rankine–Hugoniot equations, $C_1(\eta)$ can be found to be

$$C_1(\eta) = \frac{\psi(1)^{\gamma}}{f(1)} = \left[\frac{\gamma+1}{\gamma-1+2\eta}\right]^{\gamma}\frac{\gamma(\gamma+1)}{2\gamma-(\gamma-1)\eta}. \tag{8.4.13}$$

Note that, unlike the self-similar solutions for strong blast where $\theta = -(\frac{j+1}{2})$, no other integrals can be obtained from the quasi-similar equations. However, the first integral (i.e., Eq 8.4.11 or 8.4.10) provides analytical expressions for the dependent variables ψ, ϕ, and f.

Note that in the quasi-similar equations, the parameter θ is as yet not known. Thus an additional equation for θ is required. The energy integral given previously by Eq 8.2.8 is

$$1 = \frac{Iy}{\eta} - \frac{y}{\gamma(\gamma-1)(j+1)},$$

where $I = \int_0^1 (\frac{\psi\phi^2}{2} + \frac{f}{\gamma-1})\xi^j d\xi$, and the dimensionless shock radius y is given by

$$y = \left(\frac{R_s}{R_0}\right)^{j+1}.$$

An expression for $\theta(\eta)$ can be obtained by differentiating the energy integral with respect to η, that is,

$$\frac{\mathrm{d}y}{\mathrm{d}\eta}\left[\frac{I}{\eta}-\frac{1}{\gamma(\gamma-1)(j+1)}\right]+y\left[\frac{1}{\eta}\frac{\mathrm{d}I}{\mathrm{d}\eta}-\frac{I}{\eta^2}\right]=0,$$

where

$$\frac{\mathrm{d}y}{\mathrm{d}\eta}=-\left(\frac{j+1}{2}\right)\frac{y}{\theta\eta}.$$

Solving for θ, we obtain

$$\theta=\frac{\left(-\frac{j+1}{2}\right)\left(1-\frac{C\eta}{I}\right)}{\left(1-\frac{\mathrm{d}\ln I}{\mathrm{d}\ln\eta}\right)}, \tag{8.4.14}$$

where

$$C=\frac{1}{\gamma(\gamma-1)(j+1)}.$$

The quasi-similarity approximation can be used to evaluate the derivative of the integral I with respect to η. However, the quasi-similarity approximation cannot be applied to a function which consists of the sum of two other functions (i.e., $I(\eta)=I_1(\eta)+I_2(\eta)$). To evaluate the derivative of I, we first write

$$\frac{\mathrm{d}\ln I}{\mathrm{d}\ln\eta}=\frac{\eta}{I}\left(\frac{\mathrm{d}I_1}{\mathrm{d}\eta}+\frac{\mathrm{d}I_2}{\mathrm{d}\eta}\right),$$

where

$$I_1=\int_0^1\left(\frac{f}{\gamma-1}\right)\xi^j\mathrm{d}\xi,$$

$$I_2=\int_0^1\left(\frac{\psi\phi^2}{2}\right)\xi^j\mathrm{d}\xi.$$

Using the quasi-similar approximation, we now write

$$\frac{\mathrm{d}\ln I}{\mathrm{d}\ln\eta}=\frac{1}{I}\left\{G_1(\eta)\int_0^1\frac{f}{\gamma-1}\xi^j\mathrm{d}\xi+G_2(\eta)\int_0^1\frac{\psi\phi^2}{2}\xi^j\mathrm{d}\xi\right\},$$

where

$$G_1(\eta)=\frac{-(\gamma-1)\eta}{2\gamma-(\gamma-1)\eta},$$

$$G_2(\eta)=-2\eta\left(\frac{1}{(\gamma-1)+2\eta}+\frac{1}{1-\eta}\right).$$

Substituting the above equation into Eq 8.4.14, we obtain

$$\theta=-\left(\frac{j+1}{2}\right)\left(1-\frac{C\eta}{I}\right)\left\{1-\frac{1}{I}\left[G_1(\eta)\int_0^1\frac{f}{\gamma-1}\xi^j\mathrm{d}\xi+G_2(\eta)\int_0^1\frac{\psi\phi^2}{2}\xi^j\mathrm{c}\xi\right]\right\}^{-1}. \tag{8.4.15}$$

The above equation can be used to determine the correct value of θ by iteration. For example, to obtain the solution for a chosen value of the local Mach number (i.e., η), a trial value of θ is first chosen and the quasi-similar equations (i.e., Eqs 8.4.1–8.4.3) are then integrated numerically. The Rankine–Hugoniot equations are used to determine the values of $\phi(1)$, $\psi(1)$, and $f(1)$ at the shock front $\xi = 1$. From the solutions for $\phi(\xi)$, $f(\xi)$, $\psi(\xi)$, the value of the integrals I, I_1, and I_2 are then evaluated. A value of θ can then be obtained from Eq 8.4.15. The value for θ from the energy integral is then compared to the assumed value. The procedure can be repeated using another trial value of θ until the two values of θ agree within a prescribed accuracy.

Note that the solution must also satisfy the rear boundary condition at the center of symmetry, that is, zero particle velocity at $r = 0$, $\phi(0) = 0$. This rear boundary condition can also be used as a criterion for the iteration for the correct value of θ instead of using the energy integral (i.e., Eq 8.4.8). However, the profiles of $\psi(\xi)$, $\phi(\xi)$, and $f(\xi)$ of the quasi-similar solutions are not valid for the entire range of $0 \le \xi \le 1$. This is due to the fact that the quasi-similar solution does not satisfy the conservation of the total mass, that is, the mass integral. To demonstrate this, we first multiply Eq 8.4.1 by ξ^j and re-arranging yields

$$\frac{\mathrm{d}}{\mathrm{d}\xi}\left[\psi\xi^j(\phi - \xi)\right] + \psi\xi^j(j + 1 - \Lambda_1) = 0.$$

Integrating the above equation gives

$$\int_0^1 \psi\xi^j \mathrm{d}\xi = -\frac{\psi(1)\,(\phi(1) - 1)}{(j + 1) - \Lambda_1(\eta)}, \tag{8.4.16}$$

where $\psi(1)$ and $\phi(1)$ are values at the shock front $\xi = 1$ given by the Rankine–Hugoniot equations. The conservation of mass flux across the shock front gives

$$\rho_0 \dot{R}_s = \rho_1(\dot{R}_s - u_1),$$

or

$$\psi(1)\,(1 - \phi(1)) = 1,$$

where $\dot{R}_s$ is the shock velocity. Thus Eq 8.4.16 becomes

$$\int_0^1 \psi\xi^j \mathrm{d}\xi = \frac{1}{(j + 1) - \Lambda_1(\eta)}. \tag{8.4.17}$$

The mass integral can be written as

$$\int_0^{R_s} \rho_0 k_j r^j \mathrm{d}r = \rho R_s^{j+1} k_j \int_0^1 \psi\xi^j \mathrm{d}\xi = \frac{\rho_0 R_s^{j+1} k_j}{j + 1}.$$

Thus

$$\int_0^1 \psi\xi^j \mathrm{d}\xi = \frac{1}{j + 1}, \tag{8.4.18}$$

which differs from Eq 8.4.17 based on the quasi-similar approximation. For strong shocks where $\eta \to 0$, $\Lambda_1 = \frac{-4\theta\eta}{(\gamma-1)+\eta} \to 0$ and Eq 8.4.17 reduces to 8.4.18.

However, for finite shock strengths, Eqs 8.4.17 and 8.4.18 are different. The difference between them increases for decreasing shock strength M_s. For example, for $\eta = 0.25$ ($M_s = 2$), the error can be as large as 50%. On the other hand, the expression for θ from the energy integral (i.e., Eq 8.4.15) is much less sensitive to the detailed flow distribution behind the shock. Thus for the iteration for θ, the energy integral should be used.

When the value of θ for a chosen local value of the shock Mach number has been obtained, the profiles $\phi(\xi)$, $\psi(\xi)$, $f(\xi)$, and the energy integral I can be determined. Since the energy integral provides a relationship between the shock strength r and the shock radius $y = \left(\frac{R_s}{R_0}\right)^{j+1}$, the shock trajectory can be found from the definition

$$\dot{R}_s = \frac{dR_s}{dt}.$$

With $y = \left(\frac{R_s}{R_0}\right)^{j+1}$, $dR_s = \frac{R_0 dy}{y^{\frac{j}{j+1}}(j+1)}$ and since $\frac{\dot{R}_s}{c_0} = M_s = \frac{1}{\eta^{\frac{1}{2}}}$, we may write

$$\frac{c_0 t}{R_0} = \int_0^y \frac{\eta^{\frac{1}{2}} dy}{(j+1) y^{\frac{j}{j+1}}}. \tag{8.4.19}$$

Using the value of $y(\eta)$ obtained from the energy integral in the above equation yields the shock trajectory. The quasi-similar solution has been found to be quite accurate down to moderate shock strength where $M_s \sim$ o[2].

A comparison of the shock decay coefficient θ from the perturbation, quasi-similar and numerical solutions is shown in Fig. 8.4. The quasi-similar results were obtained by Lewis (1961) and the results for the perturbation solution is from Bach and Lee (1969).

Oshima (1962) also suggested an alternate method of integrating the conservation equations (Eqs 8.4.1–8.4.3) by introducing two variables $J(\xi)$ and $K(\xi)$, defined by

$$J(\xi) = \xi - \phi, \tag{8.4.20}$$

$$K(\xi) = 1 - (\xi - \phi)^2 \frac{\psi}{\gamma f}. \tag{8.4.21}$$

Using Eqs 8.4.2 and 8.4.3, one obtains

$$\phi' = \frac{\phi(\phi - \xi)\Lambda_2 - \frac{f}{\psi}\left(\Lambda_3 - \frac{\gamma j \phi}{\xi}\right)}{(\phi - \xi)^2 - \frac{\gamma f}{\psi}}.$$

Using the above equation for ϕ', differentiating Eqs 8.4.20 and 8.4.21 gives

$$\frac{dJ}{d\xi} = A + \frac{1}{K}\left(B - \frac{jJ}{\xi}\right) + \frac{C\xi}{J}\left(1 - \frac{1}{K}\right), \tag{8.4.22}$$

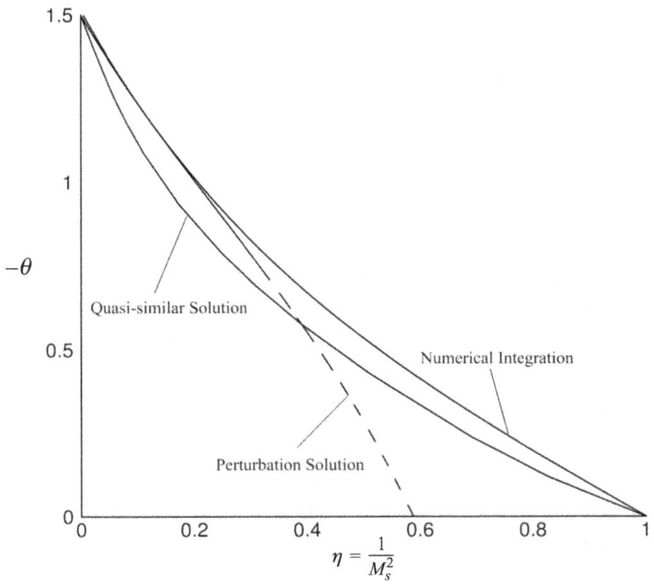

Figure 8.4 A Comparison of the Shock Decay Coefficient θ from the Perturbation, Quasi-Similar, and Numerical Solutions

where

$$A = 1 - \Lambda_2 = 1 + \frac{(1+\eta)\theta}{1-\eta},$$
$$B = \Lambda_2 - \frac{\Lambda_3}{\gamma} + j = \left[\frac{-(1+\eta)}{1-\eta} + \frac{4}{2\gamma-(\gamma-1)\eta}\right]\theta + j,$$
$$C = \Lambda_2 = \frac{-(1+\eta)}{1-\eta}\theta.$$

From Eqs 8.4.1 and 8.4.3, one gets

$$\frac{\psi'}{\psi} = \frac{1}{(\phi-\xi)}\left(\Lambda_1 - \phi' - \frac{j\phi}{\xi}\right) \qquad \frac{f'}{f} = \frac{1}{(\phi-\xi)}\left(\Lambda_3 - \gamma\phi' - \frac{\gamma j\phi}{\xi}\right).$$

Differentiating Eq 8.4.22 and using the above expressions and Eq 8.4.20, one obtains

$$\frac{\mathrm{d}K}{\mathrm{d}\xi} = \frac{1-K}{J}\left[D - (\gamma+1)\frac{\mathrm{d}J}{\mathrm{d}\xi} - \frac{(\gamma+1)jJ}{\xi}\right], \tag{8.4.23}$$

where

$$D = j(\gamma+1) + \Lambda_1 - \Lambda_3 = j(\gamma+1) + \theta\left[\frac{4\gamma}{2\gamma-(\gamma-1)\eta} - \frac{4\eta}{(\gamma-1)+2\eta}\right].$$

Equations 8.4.20 and 8.4.21 constitute a pair of equations for the variables $J(\xi)$ and $K(\xi)$ which can be integrated from the shock front $\xi = 1$ to the center of symmetry where $\xi = 0$. The boundary conditions for $J(1)$ and $K(1)$ at the shock front can be found from the Rankine–Hugoniot equations as

$$J(1) = 1 - \phi(1) = \frac{\gamma - 1 + 2\eta}{\gamma + 1}, \tag{8.4.24}$$

$$K(1) = 1 - \frac{(1 - \phi(1))^2 \psi(1)}{\gamma f(1)} = \frac{(\gamma + 1)(1 - \eta)}{2\gamma - (\gamma - 1)\eta}. \tag{8.4.25}$$

At the center of symmetry, $\xi = 0, J(0) = 0$, and $K(0) = 1$. When $J(\xi)$ and $K(\xi)$ are known, $\phi(\xi)$, $\psi(\xi)$, $J(\xi)$, and $f(\xi)$ can be obtained from the basic equations and the first integral. The procedure for the solution is to choose a value of θ and Eqs 8.4.20 and 8.4.21 are integrated numerically from the shock front $\xi = 1$ to $\xi = 0$. The desired value of θ is obtained by iteration to satisfy the rear boundary condition at $\xi = 0$, that is, $J(0) = 0$ and $K(0) = 1$. For strong shocks where $\eta \ll 1$, this method of solution gives accurate results. However, at low shock strengths, that is, $\eta \simeq 1$, the results are in error since the quasi-similar approximation does not satisfy the conservation of mass. Thus the energy integral should be used as the criterion for the iteration of the correct solution. Figure 8.4 compares the decay coefficient obtained from perturbation, quasi-similar, and numerical solutions.

In terms of the variables $J(\xi)$ and $K(\xi)$, Oshima noted that the solution is closely approximated by

$$J(\xi) = \alpha\xi + \beta\xi^n, \tag{8.4.26}$$

where α, β, and n can be obtained as

$$\alpha = \frac{A + B}{j + 1},$$
$$\beta = J(1) - \alpha,$$
$$n = \left[\left(\frac{\mathrm{d}J}{\mathrm{d}\xi}\right)_{\xi=1} - \alpha\right]\beta^{-1}.$$

A, B, and $J(1)$ are given by Eqs 8.4.22 and 8.4.24 and $\left(\frac{\mathrm{d}J}{\mathrm{d}\xi}\right)_{\xi=1}$ can be calculated from Eq 8.4.22 using the boundary conditions at $\xi = 1$. Substituting Eq 8.4.26 into the differential equations for $K(\xi)$, that is, Eq 8.4.23, we can obtain the solution for $K(\xi)$ as

$$K(\xi) = 1 - (1 - K(1))\left(\frac{J(\xi)}{J(1)}\right)^{\gamma+1+\frac{D}{\alpha(n-1)}} \xi^{j(\gamma-1)-\frac{nD}{\alpha(n-1)}}. \tag{8.4.27}$$

Once $J(\xi)$ and $K(\xi)$ are found, $\psi(\xi)$ and $f(\xi)$ can be obtained, that is,

$$\psi(\xi) = \psi(1)\left(\frac{J(\xi)}{J(1)}\right)^{\frac{\Lambda_1-(j+1)}{\alpha(n-1)}} \xi^{\frac{-n(\Lambda_1-(j+1))}{\alpha(n-1)}-j}, \tag{8.4.28}$$

$$f(\xi) = f(1)\left(\frac{J(\xi)}{J(1)}\right)^{\frac{\Lambda_3-\gamma(j+1)}{\alpha(n-1)}} \xi^{\frac{-n(\Lambda_3-\gamma(j+1))}{\alpha(n-1)}-\gamma j}. \tag{8.4.29}$$

From the definition of $J(\xi)$, we can find $\phi(\xi)$ as

$$\phi(\xi) = \xi - J(\xi) = \xi(1 - \alpha) + \beta\xi^n. \tag{8.4.30}$$

With $\psi(\xi)$, $\phi(\xi)$, $f(\xi)$ known, the integral I can be evaluated, and the shock radius $y = \left(\frac{R_s}{R_0}\right)^{j+1}$ can be obtained. With $y(\eta)$ known the shock trajectory can also be determined, that is,

$$\frac{c_0 t}{R_0} = \int_0^y \frac{\eta^{\frac{1}{2}} \mathrm{d}y}{(j+1) y^{\frac{j}{j+1}}}.$$

The perturbation solution is very accurate for strong shocks where $\eta \ll 1$ but becomes less so for moderate strength shocks. Thus quasi-similar solution is good for moderate strength shocks but not good for strong blasts where the quasi-similar approximation fails. Both methods fail for weak shock when $\eta \to 1$.

8.5 Integral Method

An integral method for finite strength blast waves was developed by Bach and Lee (Bach and Lee 1970). The method is in the same spirit as the integral method in laminar boundary layer theory, where the profile of one of the dependent variables is assumed. In the continuity equation for non-steady one-dimensional flow, there are two dependent variables: the density and the particle velocity. Thus if the profile for one variable is assumed the other is obtained by integrating the continuity equation. With the density and particle velocity profiles known, the pressure profile can then be obtained from the momentum equation. With the form for the three profiles determined, the energy integral then provides an equation for determining the shock decay law. The density profile behind a blast wave can be closely approximated by a simple power law which does not vary much with the shock Mach number when the shock wave is strong. Thus it is more appropriate to assume the profile for the density distribution and obtain the particle velocity by integrating the continuity equation. To illustrate the method, we shall consider first the case of a strong blast and we also use the dimensional form of the Euler equations.

From the self-similar solution for strong blast waves, we note that the density profile can accurately be approximated by a simple power law of the following form,

$$\rho = \rho_1 \left(\frac{r}{R_s}\right)^q, \tag{8.5.1}$$

where ρ_1 is the density at the shock front $r = R_s$. The exponent q can be obtained from the mass integral, that is,

$$\int_0^{R_s} \rho k_j r^j \mathrm{d}r = \int_0^{R_s} \rho_0 k_j r^j \mathrm{d}r = \frac{\rho_0 R_s^{j+1} k_j}{j+1},$$

and substituting the density profile given by Eq 8.5.1 into the above equation yields

$$q = \left(\frac{\rho_1}{\rho_0} - 1\right)(j+1). \tag{8.5.2}$$

For strong shocks where $\frac{\rho_1}{\rho_0} = \frac{\gamma+1}{\gamma-1}$, Eq 8.5.2 gives

$$q = \frac{2}{\gamma - 1}(j+1). \tag{8.5.3}$$

Since $\gamma \simeq O[1]$, q in general is large, that is, $q \gg 1$. For example, for $\gamma = 1.4$, $q \simeq 15$ for spherical blast waves ($j = 2$). Thus the density profile is very steep and the mass swept by the shock is concentrated near the shock front.

The continuity equation is given by

$$\frac{\partial \rho}{\partial t} + \rho\frac{\partial u}{\partial r} + u\frac{\partial \rho}{\partial r} + \frac{j\rho u}{r} = 0,$$

and substituting the density profile into the above equation yields the following ordinary differential equation for the particle velocity, that is,

$$\frac{\mathrm{d}u}{\mathrm{d}r} + (j+q)\frac{u}{r} = q\frac{\dot{R}_s}{R_s}.$$

The above equation can be arranged as follows using the integrating factors r^{j+q}, that is,

$$\frac{\mathrm{d}}{\mathrm{d}r}\left(ur^{j+q}\right) = \frac{r^{j+q}q\dot{R}_s}{R_s}.$$

Integrating the above yields the particle velocity profile as

$$\frac{u}{\dot{R}_s} = \frac{q}{q+j+1}\left(\frac{r}{R_s}\right),$$

With q given by Eq 8.5.3, the above equation becomes

$$\frac{u}{\dot{R}_s} = \frac{2}{\gamma+1}\left(\frac{r}{R_s}\right).$$

For strong shocks for which,

$$\frac{u_1}{\dot{R}_s} = \frac{2}{\gamma+1},$$

we obtain

$$u = u_1\left(\frac{r}{R_s}\right).$$

Thus for a power law density profile, the velocity profile is linear in accord with the self-similar solution for strong blasts.

The momentum equation is given by

$$\frac{\partial u}{\partial t} + u\frac{\partial u}{\partial r} + \frac{1}{\rho}\frac{\partial p}{\partial r} = 0,$$

and substituting the density and velocity profile into the above equation yields

$$\frac{\partial p}{\partial r} = \rho_1\left(\frac{r}{R_s}\right)^q\left[\frac{2}{\gamma+1}\frac{\dot{R}_s^2}{R_s^2} - \frac{2}{\gamma+1}\frac{\ddot{R}_s}{R_s}r - \left(\frac{2}{\gamma+1}\right)^2\frac{\dot{R}_s^2 r}{R_s^2}\right],$$

Integrating the above equation yields

$$p(r,t) - p(0,t) = \frac{2\rho_0 \dot{R}_s^2}{(\gamma-1)(q+2)}\left(1-\theta-\frac{2}{\gamma+1}\right)\left(\frac{r}{R_s}\right)^{q+2},$$

where $p(0,t)$ can be evaluated using the boundary condition at the shock front $r = R_s$ for which $p_1 = \rho_0 \dot{R}_s^2 \left(\frac{2}{\gamma+1}\right)$. Doing so, the pressure profile can be obtained as

$$p(r,t) = \left(\frac{2}{\gamma+1}\right)\rho_0 \dot{R}_s^2 \left[1 - \left(\frac{\gamma+1}{\gamma-1}\right)\frac{1}{(q+2)} \left(1-\theta+\frac{2}{\gamma+1}\right)\left(1-\left(\frac{r}{R_s}\right)^{q+2}\right)\right]. \quad (8.5.4)$$

In the above equation, the shock decay coefficient θ is defined by

$$\theta = \frac{R_s \ddot{R}_s}{\dot{R}_s^2}.$$

Thus, unlike the density and the velocity profile which are completely determined, the pressure profile contains the unknown shock decay coefficient θ. However, from dimensional considerations of the energy integral for a strong blast, θ can be determined as

$$\theta = -\frac{j+1}{2},$$

corresponding to a shock trajectory $R_s = At^N$ with $N = \frac{1}{1-\theta}$. The energy integral also provides the dependence of the shock strength on the shock radius, that is,

$$\dot{R}_s^2 = \left(\frac{E_0}{\rho_0 k_j I}\right)\frac{1}{R_s^{j+1}}, \quad (8.5.5)$$

where

$$I = \int_0^1 \left(\frac{\psi\phi^2}{2} + \frac{f}{\gamma-1}\right)\xi^j \mathrm{d}\xi, \quad (8.5.6)$$

and $\psi = \frac{\rho}{\rho_0}$, $\phi = \frac{u}{\dot{R}_s}$, $f = \frac{p}{\rho_0 \dot{R}_s^2}$, $\xi = \frac{r}{R_s}$. With the density, velocity, and pressure profile known (i.e., Eqs 8.5.1, 8.5.3, and 8.5.4), the integral I can be found, and thus the complete solution is determined. The approximate solution described above is remarkably accurate when compared to the exact self-similar solution for strong blast waves.

The methodology described above for strong shocks can readily be extended to finite strength shock waves. Using now the non-dimensional equations in blast wave coordinates, the density profile is written as

$$\psi(\xi,\eta) = \psi(1,\eta)\xi^{q(\eta)}, \quad (8.5.7)$$

where the exponent q from the mass integral is obtained as

$$q(\eta) = (j+1)\left(\psi(1,\eta)-1\right), \quad (8.5.8)$$

where $\psi(1,\eta)$ is given by the Rankine–Hugoniot equations as

$$\psi(1,\eta) = \frac{\gamma+1}{(\gamma-1)+2\eta}. \tag{8.5.9}$$

Substituting the density profile into the continuity equation in blast wave coordinates below, that is,

$$(\phi-\xi)\frac{\partial\psi}{\partial\xi} + \psi\frac{\partial\phi}{\partial\xi} + \frac{j\phi\psi}{\xi} = 2\theta\eta\frac{\partial\psi}{\partial\eta},$$

we obtain

$$\frac{\partial\phi}{\partial\xi} + (j+q)\frac{\phi}{\xi} = q + \frac{2\theta\eta}{\psi(1,\eta)}(1+(j+1)\psi(1,\eta)\ln\xi)\frac{\mathrm{d}\psi(1,\eta)}{\mathrm{d}\eta}.$$

The above equation can be integrated to yield

$$\phi(\xi,\eta) = \phi(1,\eta)\xi(1-\Theta\ln\xi), \tag{8.5.10}$$

where

$$\Theta = \frac{-2\theta\eta}{\phi(1,\eta)\psi(1,\eta)}\frac{\mathrm{d}\psi(1,\eta)}{\mathrm{d}\eta}.$$

For strong shocks where $M_s \to \infty$ and $\eta \to 0$, Eq 8.5.10 reduces to the linear velocity profile given previously by Eq 8.5.3. Note that, unlike the solution for strong shocks, the velocity profile (Eq 8.5.10) now contains the parameter $\theta(\eta)$ which is, as yet, not known.

Substituting the density and velocity profiles into the momentum equation below,

$$(\phi-\xi)\frac{\partial\phi}{\partial\xi} + \theta\phi + \frac{1}{\psi}\frac{\partial f}{\partial\xi} = 2\theta\eta\frac{\partial\phi}{\partial\eta}.$$

The pressure profile can be obtained as

$$f(\eta,\xi) = \int\Bigg\{-2\theta\eta\xi\bigg(\frac{\mathrm{d}\phi(1,\eta)}{\mathrm{d}\eta} - \frac{\mathrm{d}}{\mathrm{d}\eta}(\phi(1,\eta)\Theta)\ln\xi$$
$$+(\phi-\xi)\phi(1,\eta)(1-\Theta(1+\ln\xi))\bigg) + \theta\phi\Bigg\}\psi(1,\eta)\xi^{q(\eta)}\mathrm{d}\xi + C(\eta),$$

where $C(\eta)$ is a constant of integration and can be evaluated using the boundary condition at the shock front $\xi = 1$, that is,

$$f(1,\eta) = \frac{2}{\gamma+1} - \frac{\gamma-1}{\gamma(\gamma+1)}\eta. \tag{8.5.11}$$

Carrying out the integration and evaluating the constant of integration, the pressure profile can be written as

$$f(\xi,\eta) = f(1,\eta) + f_2\left(\xi^{q+2}-1\right) + f_3\left\{\xi^{q+2}[(q+2)\ln\xi - 1] + 1\right\}$$
$$+ f_4\left\{2 - \xi^{q+2}\left[(q+2)^2\ln\xi^2 - 2(q+2)\ln\xi + 2\right]\right\}, \tag{8.5.12}$$

where $f(1,\eta)$ is the boundary condition at the shock given by Eq 8.5.11, and the functions f_2, f_3, and f_4 are defined below

$$
\begin{aligned}
f_2 &= \frac{\psi(1,\eta)}{q+2}\Bigg[(1-\Theta)\left\{\phi(1,\eta)-\phi^2(1,\eta)\right\} \\
&\quad -\theta\left\{\phi(1,\eta)-2\eta\frac{\mathrm{d}\phi(1,\eta)}{\mathrm{d}\eta}\right\}\Bigg], \\
f_3 &= \frac{\psi(1,\eta)}{(q+2)^2}\Bigg[\theta\left\{\Theta\phi(1,\eta)-2\eta\frac{\mathrm{d}}{\mathrm{d}\eta}[\Theta\phi(1,\eta)]\right\} \\
&\quad -\Theta\phi(1,\eta)-\Theta^2\phi^2(1,\eta)+2\Theta\phi^2(1,\eta)\Bigg], \\
f_4 &= \frac{\Theta^2\phi^2(1,\eta)\psi(1,\eta)}{(q+2)^3}.
\end{aligned}
$$

The density and velocity at the shock front $\xi = 1$ are given by the Rankine–Hugoniot equations,

$$
\psi(1,\eta) = \frac{\gamma+1}{\gamma-1+2\eta}, \qquad \phi(1,\eta) = \frac{2}{\gamma+1}(1-\eta).
$$

For strong shocks where $\eta = 0$, the pressure profile reduces to

$$
f(\xi) = f(1,0) + \frac{\psi(1,0)\phi(1,0)}{q+2}\left(1-\phi(1,0)-\theta\right)\left(\xi^{q+2}-1\right),
$$

with

$$
\psi(1,0) = \frac{\gamma+1}{\gamma-1}, \qquad \phi(1,0) = \frac{2}{\gamma+1} = f(1,0).
$$

In dimensional form the above equation is identical to Eq 8.5.4 obtained previously.

To complete the solution, we need to find $\theta(\eta)$ and this is obtained from the energy integral which is given previously as

$$
1 = y\left(\frac{I}{\eta} - \frac{1}{\gamma(\gamma-1)(j+1)}\right),
$$

with

$$
\begin{aligned}
I &= \int_0^1 \left(\frac{f}{\gamma-1} + \frac{\psi\phi^2}{2}\right)\xi^j \mathrm{d}\xi, \\
y &= \left(\frac{R_s}{R_0}\right)^{j+1}, \\
R_0 &= \left(\frac{E_0}{k_j\rho_0 c_0^2}\right).
\end{aligned}
$$

Substituting the profiles for $\psi(\xi,\eta)$, $\phi(\xi,\eta)$, and $f(\xi,\eta)$ and solving for $\frac{d\theta}{d\eta}$, we obtain

$$\frac{d\theta}{d\eta} = -\frac{1}{2\eta}\left\{(\theta + 1 - 2\phi(1,\eta) - \frac{(D_1+4\eta)}{\gamma+1} - (\gamma-1)(j+1)\right.$$
$$\left.\left[\phi(1,\eta) - \frac{(D_1+4\eta)^2}{4\theta y(\gamma+1)}\right]\right\} + \frac{(D_1+4\eta)}{8\eta^2(\gamma+1)}$$
$$\left[\frac{(D_1+4\eta)\phi(1,\eta)}{\theta} - \frac{\phi(1,\eta)(\gamma+1)}{\theta\psi(1,\eta)} + 2(\eta+1) + \frac{(\gamma-1)(j+1)(\gamma+1)}{2\theta}\phi^2(1,\eta)\right]$$
$$+ 2\theta\frac{2+(\gamma-1)(j+1)}{D_1+4\eta}, \tag{8.5.13}$$

where $D_1 = \gamma(j+3) + (j-1)$. In Eq 8.5.13, the shock radius y appears. Thus an additional equation for y is required. This equation can be obtained from the definition of $y = \left(\frac{R_s}{R_0}\right)^{j+1}$. Using the identity $R_s\frac{d}{dR_s} = -2\theta\eta\frac{d}{d\eta}$, we obtain

$$\frac{dy}{d\eta} = -\left(\frac{j+1}{2}\right)\frac{y}{\theta\eta}, \tag{8.5.14}$$

Equations 8.5.13 and 8.5.14 are integrated simultaneously with the initial condition $\theta(0) = \theta_0 = -\frac{j+1}{2}$ and $y(0) = 0$. However, we note that both $\frac{d\theta}{d\eta}$ and $\frac{dy}{d\eta}$ are singular at $\eta = 0$. Thus we must seek solutions for $\theta(\eta)$ and $y(\eta)$ in the neighborhood of $\eta = 0$, that is,

$$\theta(\eta) = \theta_0 + \theta_1\eta + \theta_2\eta^2 + \cdots$$
$$y(\eta) = y_1\eta + y_2\eta^2 + \cdots$$

Substituting the expressions above into the differential equations for $\theta(\eta)$ and $y(\eta)$, the coefficients $\theta_0, \theta_1, \theta_2, y_1, y_2$, etc., can be obtained as

$$\theta_0 = -\left(\frac{j+1}{2}\right), \qquad \theta_1 = \frac{\theta_0^2}{C_1}\left[A_2 + B_1\theta_0 + \frac{C_2}{\theta_0} + \frac{E_2}{y_1\theta_0}\right],$$

$$\theta_2 = \theta_0^2\left[A_3 + (B_1-1)\theta_1 + B_2\theta_0 + \frac{1}{\theta_0}\left(\frac{C_1\theta_1^2}{\theta_0^2} - \frac{C_2\theta_1}{\theta_0} + C_3\right) + \frac{E_3}{y_1\theta_0}\right]\Big/\left(C_1 + \frac{E_1}{2y_1}\right),$$

$$\theta_3 = \theta_0^2\left[(B_1-2)\theta_2 + B_2\theta_1 + B_3\theta_0\right.$$
$$+\left\{C_1\left[\frac{2\theta_1\theta_2}{\theta_0^2} - \frac{\theta_1^3}{\theta_0^3}\right] + C_2\left(\frac{\theta_1^2}{\theta_0^2} - \frac{\theta_2}{\theta_0}\right) - \frac{C_3\theta_1}{\theta_0} + C_4\right\}\Big/\theta_0$$
$$\left.+\left\{\frac{E_1\theta_1\theta_2}{3\theta_0^2} - \frac{E_2\theta_2}{2\theta_0}\right\}\Big/y_1\theta_0\right]\Big/\left(C_1 + \frac{2E_1}{3y_1}\right),$$

and

$$y_1 = \frac{-E_1}{C_1 + A_1\theta_0}, \qquad y_2 = \frac{y_1\theta_1}{\theta_0}, \qquad y_3 = y_1\left[\frac{\theta_1^2}{\theta_0^2} - \frac{\theta_2}{2\theta_0}\right],$$

$$y_4 = y_1\left\{\frac{7\theta_1\theta_2}{2\theta_0^2} - \frac{3\theta_1^3}{\theta_0^3} - \frac{\theta_3}{\theta_0}\right\}\Big/3.$$

In the above expressions, the A's, B's, C's, etc. are given by

$$A_1 = \frac{D_1}{4(\gamma+1)}, \qquad A_2 = -\frac{1}{2} + \frac{[2+(\gamma-1)(j+1)]}{(\gamma+1)} + \frac{3D_1+4}{4(\gamma+1)},$$

$$A_3 = \frac{[1-(\gamma-1)(j+1)]}{\gamma+1},$$

$$B_1 = -\frac{1}{2}, \qquad B_2 = \frac{2[2+(\gamma-1)(j+1)]}{D_1},$$

$$B_3 = -\frac{8[2+(\gamma-1)(j+1)]}{D_1^2}.$$

$$C_1 = \frac{D_1[D_1 + j(\gamma+1)]}{4(\gamma+1)^2},$$

$$C_2 = \frac{\{-D_1^2 + D_1[6-(\gamma-1)(2j+1)] + 4j(\gamma-1)\}}{4(\gamma+1)^2},$$

$$C_3 = \frac{\{D_1[-6+(\gamma-1)(j+1)] + 4[2-(\gamma-1)(2j+1)]\}}{4(\gamma+1)^2},$$

$$C_4 = -\frac{[2-(\gamma-1)(j+1)]}{(\gamma+1)^2}.$$

$$D_1 = \gamma(j+3) + (j-1).$$

$$E_1 = -\frac{[(\gamma-1)(j+1)D_1^2]}{8(\gamma+1)},$$

$$E_2 = -\frac{(\gamma-1)(j+1)D_1}{(\gamma+1)},$$

$$E_3 = -\frac{2(\gamma-1)(j+1)}{\gamma+1}.$$

Given values for γ and j, the perturbation coefficients $\theta_0, \theta_1, \theta_2, \ldots, y_1, y_2, y_3, \ldots$ can be computed from the expressions given above. Thus, the starting values for the integration of Eqs 8.5.13 and 8.5.14 can be obtained.

The shock trajectory $R_s(t)$ can be obtained from the definition

$$\dot{R}_s = \frac{\mathrm{d}R_s}{\mathrm{d}t},$$

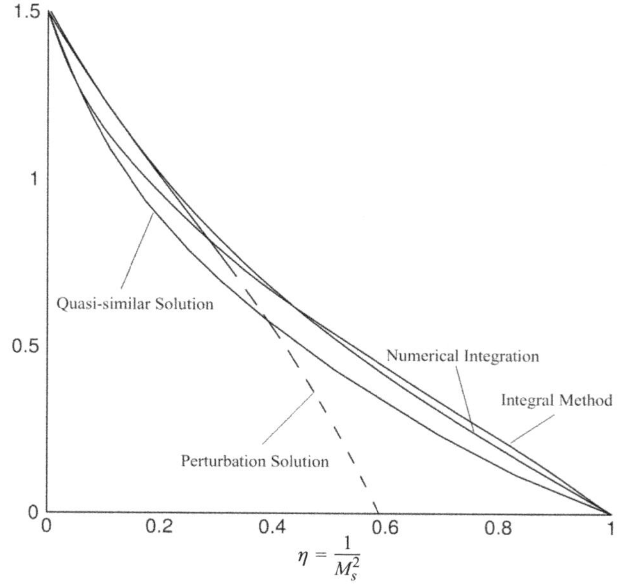

Figure 8.5 A Comparison of the Shock Decay Coefficient θ

which can be written in terms of y and η as

$$\frac{c_0 t}{R_0} = -\frac{1}{2}\int_0^{\eta} \frac{y^{\frac{1}{j+1}}\,\mathrm{d}\eta}{\theta\eta^{\frac{1}{2}}}. \tag{8.5.15}$$

The above equation is integrated simultaneously with Eqs 8.5.13 and 8.5.14. For small values of η, the shock trajectory can be obtained as

$$\frac{c_0 t}{R_0} = 2y_1^{\frac{1}{j+1}}\eta^{\frac{j+3}{2j+1}}\left\{\frac{1}{j+3} + \frac{T_1\eta}{3j+5} + \frac{T_2\eta^2}{5j+7} + \frac{T_3\eta^3}{7j+9} + \cdots\right\}. \tag{8.5.16}$$

The coefficients in the above expression are given by

$$T_1 = \frac{y_2(j+2)}{y_1(j+1)},$$

$$T_2 = (2j+3)\left\{\frac{y_3}{y_1} - \frac{jy_2^2}{2(j+1)y_1^2}\right\}\Big/(j+1),$$

$$T_3 = \left(\frac{3j+4}{j+1}\right)\left\{\frac{y_4}{y_1} - \frac{jy_2y_3}{y_1^2(j+1)} + \frac{j(2j+1)y_2^3}{6y_1^3(j+1)^2}\right\},$$

where the y's are as given previously.

Although the integral method described in this section appears to be complicated, it is in fact quite straight-forward since the long algebraic expressions need to be determined only once. The integral method satisfies the conservation integrals, that is, mass and energy integrals. The differential equations for mass and momentum are also satisfied. However, the differential equation for the conservation of energy is not satisfied

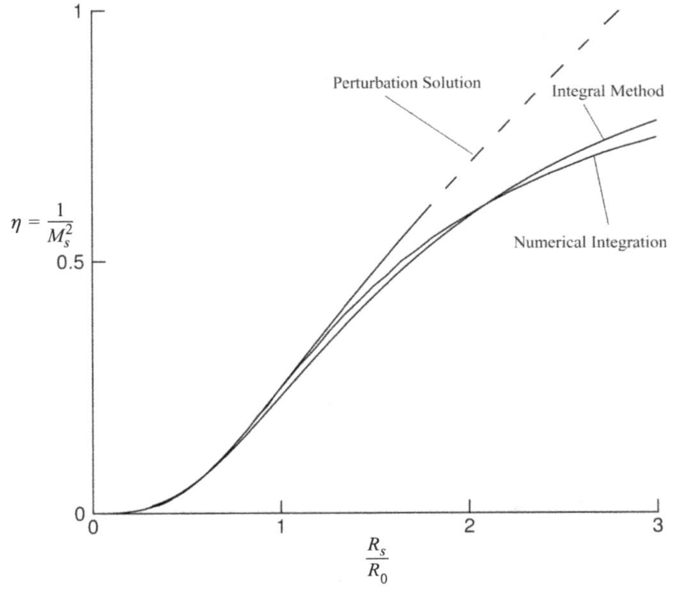

Figure 8.6 A Comparison of the Variation of the Shock Strength

although the energy integral is satisfied. It is well known that blast wave decay depends on the averaged energy density behind the shock, and is not sensitive to the details of the energy profile. As long as the global conservation of energy (that is, energy integral) is satisfied, the blast wave decay could be determined.

A comparison of the shock decay coefficient with various non-similar methods and numerical solution is illustrated in Fig. 8.5 for a spherical blast ($\gamma = 1.4$). The variation of the shock strength η with shock radius $\frac{R_s}{R_0}$ is shown in Fig. 8.6.

9 Implosions

9.1 Introduction

In an explosion, the blast wave expands and decays as the shocked volume increases and the energy density decreases. In an implosion, the shock wave converges and the shock amplifies. Self-similar solutions exist for strong converging shock waves when the shock radius becomes small compared to the initial radius of the shock. The validity of the self-similar solution is also limited to a small region of flow behind the shock since the rest of the flow field retains the initial non-similar motion. Thus the solution for implosion is an asymptotic self-similar solution when the initial non-similar motion is "forgotten" as the shock approaches the center of symmetry. The propagation of a shock wave in a decreasing density medium is similar to the implosion problem. There, the amplification of the shock is a result of the increase in the energy density due to decreasing mass ahead of the shock. The self-similar solution for converging shock waves differs from that for explosions in that the time exponent of the shock trajectory cannot be determined a priori from dimensional considerations of the energy integral. The time exponent is determined by seeking a solution that is continuous at the singularities in the flow field. Self-similar solutions of this type are referred to as "similarity solutions of the second class" by Zel'dovich and Raizer (1966).

9.2 Implosions

Consider a spherical or cylindrical shock wave initiated at some initial radius R_i. When the shock radius $R_s \ll R_i$, there is no longer a characteristic length scale in the problem and hence the motion becomes self similar. We also require that the flow field be limited to a radius $r \simeq R_s$ which is small compared to the initial radius R_i.

Assuming a strong shock, the basic conservation equation in non-dimensional blast wave variables ϕ, f, ψ, ξ can be written as

$$(\phi - \xi)\psi' + \psi\phi' + \frac{j\phi\psi}{\xi} = 0, \tag{9.2.1}$$

$$(\phi - \xi)\phi' + \theta\phi + \frac{1}{\psi}f' = 0, \tag{9.2.2}$$

$$(\phi - \xi)f' + \gamma f\phi' + \left(2\theta + \frac{\gamma j\phi}{\xi}\right)f = 0, \tag{9.2.3}$$

where $\psi = \frac{\rho}{\rho_0}$, $\phi = \frac{u}{\dot{R}_s}$, $f = \frac{p}{\rho_0 \dot{R}_s^2}$ and the similarity variable $\xi = \frac{r}{R_s}$. The prime superscript denotes differentiation with respect to ξ. The parameter θ in the above equations is defined as

$$\theta = \frac{R_s \ddot{R}_s}{\dot{R}_s^2}, \tag{9.2.4}$$

and, for a self-similar solution to be valid, $\theta =$ constant. Integrating Eq 9.2.4 gives

$$\dot{R}_s = -C R_s^{\theta}, \tag{9.2.5}$$

where the negative sign is chosen for implosion since $\dot{R}_s$ increases for decreasing shock radius R_s. Integrating Eq 9.2.5 gives the shock trajectory

$$R_s = A\left(-t^N\right), \tag{9.2.6}$$

with

$$N = \frac{1}{1-\theta} \quad \text{or} \quad \theta = -\left(\frac{1-N}{N}\right). \tag{9.2.7}$$

In Eq 9.2.6, we have chosen $R_s = 0$ when $t = 0$. Thus the implosion process occurs at negative times and increases to zero as the shock reaches the center of symmetry.

Solving for the derivatives ψ', ϕ', f' from Eqs 9.2.1–9.2.3, we write

$$\phi' = \frac{\left(2\theta f + \frac{\gamma j f \phi}{\xi}\right)\frac{1}{\psi} - \theta\phi(\phi - \xi)}{(\phi - \xi)^2 - \frac{\gamma f}{\psi}}, \tag{9.2.8}$$

$$f' = -\psi\left(\theta\phi + (\phi - \xi)\phi'\right), \tag{9.2.9}$$

$$\psi' = -\frac{\left(\psi\phi' + \frac{j\phi\psi}{\xi}\right)}{(\phi - \xi)}, \tag{9.2.10}$$

where for convenience we have expressed f' and ψ' in terms of ϕ'. Prior to the integration of the similarity equations (i.e., Eqs 9.2.8–9.2.10) the value of θ must be known.

Unlike the explosion problem, the energy integral cannot be used since it extends to the non-similar region where the initial conditions dominate and the self-similar solution is not valid there.

To determine the value of θ, we must examine the self-similar flow field described by Eqs 9.2.8–9.2.10. We note that there exists three singularities $\xi = 0$, $(\phi - \xi) = 0$, and $(\phi - \xi)^2 - \frac{\gamma f}{\psi} = 0$. $\xi = 0$ corresponds to the center of symmetry, $(\phi - \xi) = 0$ corresponds to an interface or piston surface. $(\phi - \xi)^2 - \frac{\gamma f}{\psi} = 0$ is when a constant ξ line coincdes with one of the physical characteristics $\frac{dr}{dt} = u \pm c$ and is generally referred to as sonic singularity. It is the relevant singularity we must consider in the implosion problem. To demonstrate this, we note that for a constant ξ line, that is,

$$\xi = \frac{r}{R_s} = \text{constant},$$

thus $\frac{dr}{dt} = \dot{R}_s \xi$ or $\xi = \frac{1}{\dot{R}_s}\frac{dr}{dt}$.

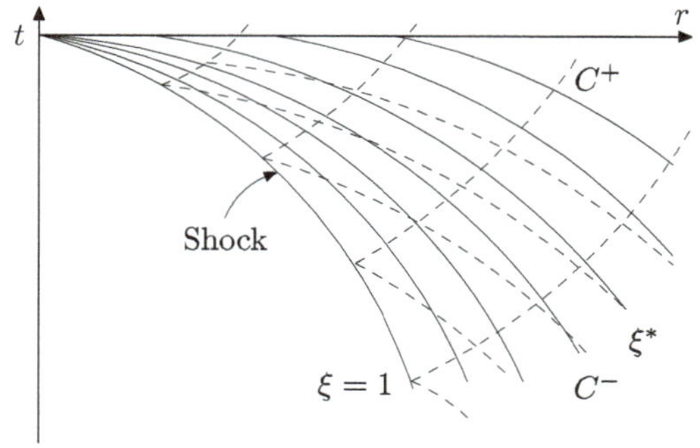

Figure 9.1 Physical Characteristics and Constant ξ Lines

The singularity $(\phi-\xi)^2-\frac{\gamma f}{\psi}=0$ gives $\xi=\phi\pm\sqrt{\frac{\gamma f}{\psi}}$, which in dimensional variables give $\frac{dr}{dt}=u\pm c$ since $c^2=\frac{\gamma p}{\rho}$. Thus the singularity $(\phi-\xi)^2-\frac{\gamma f}{\psi}=0$ corresponds to the coincidence of a constant ξ line with the physical characteristics $\frac{dr}{dt}=u\pm c$. For the case of implosion, this is illustrated in Fig. 9.1.

In the flow region bounded by $1\le\xi\le\xi^*$, the C^- characteristic $\frac{dr}{dt}=-u-c=-(u+c)$ catches up with the shock front $\xi=1$ and reflects off as a C^+: $\frac{dr}{dt}=-u+c=-(u-c)$. The singularity at $\xi=\xi^*$ where $(\phi^*-\xi^*)^2-\frac{\gamma f^*}{\psi^*}=0$ coincides with a C^- characteristic which arrives at $r=0$ at the same time as the shock as shown in Fig. 9.1. In the region $\xi>\xi^*$, disturbances cannot influence the shock motion and thus the flow there may be non-similar. The self-similar region is bounded by $1\le\xi\le\xi^*$. Since information from the non-self-similar region $\xi>\xi^*$ cannot transmit to the self-similar region $\xi<\xi^*$, it seems logical that the solutions for the two regions $1\le\xi\le\xi^*$ and $\xi>\xi^*$ must match along their common boundary $\xi=\xi^*$. From physical considerations, the criterion for determining an acceptable solution is to guarantee the matching of the solutions at $\xi=\xi^*$. In other words, the value of θ (or equivalently the time exponent N) must be so chosen that the solution is continuous across the ξ^* boundary. It should be noted that the solution determined in this manner is independent of the initial conditions of the problem. The existence of the ξ^* boundary essentially isolates the region behind the shock from the rest of the flow field. Thus self-similar solutions can evolve asymptotically from some non-similar initial flow.

From Eq 9.2.8, we see that when the denominator $(\phi^*-\xi^*)^2-\frac{\gamma f^*}{\psi^*}\to 0$, we must require the numerator to vanish simultaneously. Thus

$$\left(2\theta f^*+\frac{\gamma j f^*\phi}{\xi^*}\right)\frac{1}{\psi^*}-\theta\phi^*(\phi^*-\xi^*)\to 0,$$

and solving for θ, we get

$$\theta=\frac{\frac{\gamma j f^*\phi^*}{\xi^*\psi^*}}{\phi^*(\phi^*-\xi^*)-\frac{2f^*}{\psi^*}}, \qquad (9.2.11)$$

where the asterisk superscript represents the condition at the singularity $\xi=\xi^*$. Note that, if Eq 9.2.11 is satisfied, f' and ψ' will also be finite at $\xi=\xi^*$. Thus, the method

Table 9.1 Spherical Shocks

γ	θ	N	N(Butler)	N(Welsh)
1.2	−0.320757	0.757142	0.757142	0.757142
1.4	−0.394361	0.717174	0.717173	0.717174
5/3	−0.452692	0.688377	0.688377	0.688377
3	−0.571314	0.636410	—	0.636411

of solution for the implosion problem consists of integrating the similarity equations at the shock front, $\xi = 1$, using the strong shock condition

$$\psi(1) = \frac{\gamma+1}{\gamma-1}, \qquad \phi(1) = f(1) = \frac{2}{\gamma+1}, \tag{9.2.12}$$

using a trial value of θ. The correct solution is obtained when the value of θ satisfies the regularity condition given by Eq 9.2.11 at the singularity ξ^*. The values for θ (and the time exponent N) for spherical and cylindrical converging strong shock waves for various values of γ are given in Tables 9.1 and 9.2. For comparison, the values obtained by Butler (1954) and Welsh (1966) are also shown.

9.3 Solution in the State Plane

Instead of the simultaneous integration of the three similarity equations (i.e., Eqs 9.2.8–9.2.10) for the variables ψ, ϕ, and f, it is more convenient to combine the three equations to obtain a single equation using state variables Z and V defined as

$$Z = \frac{N^2}{\xi^2}\frac{\gamma f}{\psi}, \tag{9.3.1}$$

$$V = \frac{N\phi}{\xi}. \tag{9.3.2}$$

Since $c^2 = \frac{\gamma p}{\rho} = \dot{R}_s^2\left(\frac{\gamma f}{\psi}\right)$ and $u = \dot{R}_s\phi$, the variables Z and V denote the sound speed and the particle velocity, which are the state variables of the gas. From Eqs 9.3.1 and 9.3.2, we write

$$\frac{dZ}{Z} = \frac{df}{f} - \frac{d\psi}{\psi} - 2\frac{d\xi}{\xi}, \tag{9.3.3}$$

$$\frac{dV}{V} = \frac{d\phi}{\phi} - \frac{d\xi}{\xi}. \tag{9.3.4}$$

Table 9.2 Cylindrical Shocks

γ	θ	N	N(Butler)	N(Welsh)
1.2	−0.161221	0.861163	0.861163	0.861163
1.4	−0.197143	0.835322	0.835217	0.835323
5/3	−0.226054	0.815625	0.815625	0.688377
3	−0.289214	0.775666	—	0.775667

Using Eqs 9.3.1–9.3.4, the similarity equations (i.e., Eqs 9.2.1–9.2.3) can be written as

$$(V-N))\frac{\mathrm{d}\ln\psi}{\mathrm{d}\ln V}+(j+1)\frac{V\mathrm{d}\ln\xi}{\mathrm{d}\ln V}+V=0, \tag{9.3.5}$$

$$\frac{Z}{V}\frac{\mathrm{d}\ln\psi}{\mathrm{d}\ln V}+\left(\frac{2Z}{\gamma}+V(V-1)\right)\frac{\mathrm{d}\ln\xi}{\mathrm{d}\ln v}+\frac{Z}{\gamma}\frac{\mathrm{d}\ln Z}{\mathrm{d}\ln V}+V(V-N)=0, \tag{9.3.6}$$

$$(V-N))\frac{\mathrm{d}\ln\psi}{\mathrm{d}\ln V}+[2(V-1)+\gamma V(j+1)]\frac{\mathrm{d}\ln\xi}{\mathrm{d}\ln V}+(V-N)\frac{\mathrm{d}\ln Z}{\mathrm{d}\ln V}+\gamma V=0. \tag{9.3.7}$$

Solving for the derivatives $\frac{\mathrm{d}\ln Z}{\mathrm{d}\ln V}$, $\frac{\mathrm{d}\ln\psi}{\mathrm{d}\ln V}$, and $\frac{\mathrm{d}\ln\xi}{\mathrm{d}\ln V}$ from the above equations, we get

$$\frac{\frac{(V-N)}{N}}{V}\frac{\mathrm{d}\ln Z}{\mathrm{d}\ln V}=\frac{\left[(V-N)^2-Z\right][2-(2+(j+1)(\gamma-1)V)]}{Z\left[V(j+1)-\frac{2}{\gamma}(1-N)\right]-V(V-1)(V-N)}-(\gamma-1), \tag{9.3.8}$$

$$\frac{\mathrm{d}\ln\psi}{\mathrm{d}\ln V}=\frac{-V}{(V-N)}\left[\frac{V(V-N)(V-1)-(j+1)V(V-N)^2+\frac{Z2}{\gamma}(1-N)}{V(V-1)(V-N)-Z\left[V(j+1)-\frac{2}{\gamma}(1-N)\right]}\right], \tag{9.3.9}$$

$$\frac{\mathrm{d}\ln\xi}{\mathrm{d}\ln V}=\frac{V\left(Z-(V-N)^2\right)}{V(V-1)(V-N)-Z\left[(j+1)V-\frac{2}{\gamma}(1-N)\right]}. \tag{9.3.10}$$

From Eqs 9.3.9 and 9.3.10, we may eliminate $\mathrm{d}\ln V$ and write

$$\frac{\mathrm{d}\ln\psi(V-N)}{\mathrm{d}\ln\xi}=\frac{-V(j+1)}{V-N}. \tag{9.3.11}$$

Instead of integrating simultaneously Eqs 9.2.8–9.2.10, we may now integrate Eq 9.3.8 and once the solution for $Z(V)$ is obtained, then $\psi(\xi)$ and $V(\xi)$ can be found from Eqs 9.3.10 and 9.3.11. The pressure distribution $f(\xi)$ and the particle velocity $\phi(\xi)$ can be obtained from the definition of Z and V (i.e., Eqs 9.3.1 and 9.3.2). It should be noted that a single integral curve $Z(V)$ represents the complete solution. Thus it is sometimes more convenient to study the mathematical properties of a solution if we use the $(Z,\ V)$ coordinates.

The singularities of the similarity equations correspond to the following in Z, V coordinates, that is,

$$\xi=0, \tag{9.3.12}$$

$$(\phi-\xi)=0 \quad\rightarrow\quad (V-N)=0, \tag{9.3.13}$$

$$(\phi-\xi)^2-\frac{\gamma f}{\psi}=0 \quad\rightarrow\quad (V-N)^2-Z=0. \tag{9.3.14}$$

The strong shock conditions are given by

$$V(1)=\frac{2N}{\gamma+1}, \tag{9.3.15}$$

$$Z(1)=\frac{2\gamma N^2(\gamma-1)}{(\gamma+1)^2}=\gamma V(1)(N-V(1)). \tag{9.3.16}$$

Thus the strong shock condition is represented by the parabola given by Eq 9.3.16.

The sonic singularity is represented by the parabola $Z = (V - N)^2$. The regularity condition given by Eq 9.2.11 can be expressed in (Z, V) coordinates as

$$Z = \frac{V(V-1)(V-N)}{(j+1)V - \frac{2}{\gamma}(1-N)}. \tag{9.3.17}$$

Equating the value of Z in Eqs 9.3.14 and 9.3.17 gives a quadratic equation for V whose solution is given by

$$V = \frac{\alpha \pm \sqrt{\alpha^2 - 4\beta}}{2}, \tag{9.3.18}$$

where

$$\alpha = \frac{(j+1)N + \frac{2}{\gamma}(1-N) - 1}{j},$$

$$\beta = \frac{2N(1-N)}{\gamma j}.$$

The integral curve for converging shocks starts at the shock front $\xi = 1$

$$Z_1 = \frac{2\gamma N^2(\gamma - 1)}{(\gamma+1)^2},$$

and

$$V_1 = \frac{2N}{\gamma + 1}, \tag{9.3.19}$$

and terminates at $\xi \to 0$ where $Z \to 0$, $V \to 0$. The correct value of the time exponent N must be such that the regularity condition Eq 9.3.17 is satisfied at the singularity (Eq 9.3.14), that is, at the intersection of the two parabolas given by Eqs 9.3.14 and 9.3.17. The two roots of Eq 9.3.18 correspond to the intersection of the parabolas representing the singularity and the regularity conditions (Eqs 9.3.14 and 9.3.17). A sketch of the implosion solution in the $Z(V)$ plane is shown in Fig. 9.2. The correct solution is obtained by iteration for the value of N such that the integral curve passes through the intersection of the singularity and regularity parabolas shown in Fig. 9.2.

9.4 Shock Propagation in a Non-uniform Density Medium

We consider the density ρ_0 to vary as a power law of distance from the origin, that is, $\rho_0 = Bx^\delta$ where B and δ are positive constants. We assume the shock to start at some initial value of X_0 and propagate toward $x = 0$, as illustrated in Fig. 9.3.

The basic conservation equations in planar geometry are given by

$$\frac{\partial \rho}{\partial t} + \frac{\partial(\rho u)}{\partial x} = 0, \tag{9.4.1}$$

$$\frac{\partial u}{\partial t} + u\frac{\partial u}{\partial x} + \frac{1}{\rho}\frac{\partial p}{\partial x} = 0, \tag{9.4.2}$$

$$\left(\frac{\partial}{\partial t} + u\frac{\partial}{\partial x}\right)\frac{p}{\rho^\gamma} = 0. \tag{9.4.3}$$

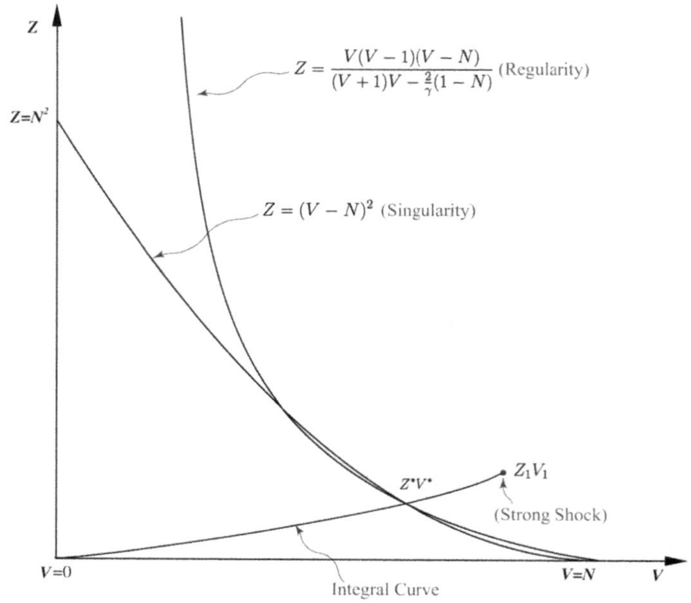

Figure 9.2 Implosion in Z–V Plane

When the shock is near the center, that is, $X_s \ll X_0$ and for a region $x \ll X_0$, there are no characteristic length scales in the problem, and self-similar solution exists if the shock wave is strong. Using the continuity equation (Eq 9.4.1) to eliminate the density in the energy equation (Eq 9.4.3), we obtain

$$\left(\frac{\partial}{\partial t} + u\frac{\partial}{\partial x}\right)p + \gamma p\frac{\partial u}{\partial x} = 0. \tag{9.4.4}$$

Defining the following non-dimensional variables as we have done earlier

$$\left.\begin{aligned} \rho &= \rho_0\psi(\xi) \\ u &= \dot{X}_s\phi(\xi) \\ p &= \rho_0\dot{X}_s^2 f(\xi) \end{aligned}\right\} \tag{9.4.5}$$

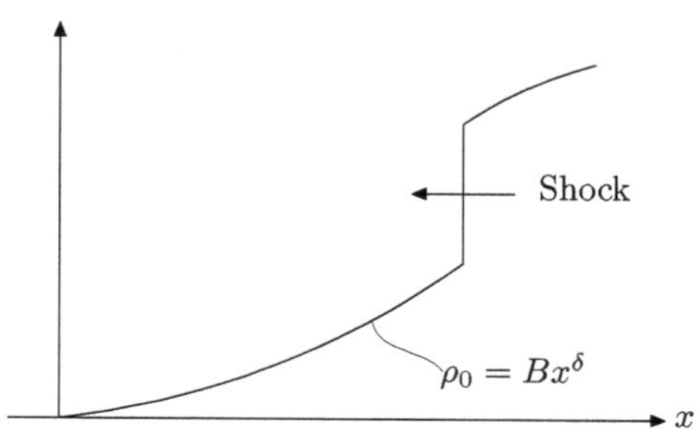

Figure 9.3 Shock Propagating Toward $x = 0$

where $\xi = \frac{x}{X_s}$, Eqs 9.4.1, 9.4.2, and 9.4.4 transform to the following

$$(\phi - \xi)\frac{d\psi}{d\xi} + \psi\frac{d\phi}{d\xi} + \delta\psi = 0, \tag{9.4.6}$$

$$(\phi - \xi)\frac{d\phi}{d\xi} + \frac{1}{\psi}\frac{df}{d\xi} + \theta\phi = 0, \tag{9.4.7}$$

$$(\phi - \xi)\frac{df}{d\xi} + \gamma f\frac{d\phi}{d\xi} + (2\theta + \delta)f = 0, \tag{9.4.8}$$

where $\theta = \frac{X_s\ddot{X}_s}{\dot{X}_s^2}$. We have also referenced the initial density distribution with respect to the shock position, that is, $\rho_0 = BX_s^{\delta}$.

For self-similar solution, θ has to be a constant and solving for the shock velocity, we get

$$\dot{X}_s = -CX_s^{\theta}, \tag{9.4.9}$$

where we have taken a negative sign since $\dot{X}_s$ is negative when the shock advances in the direction of decreasing X_s. Integrating Eq 9.4.9 gives the shock trajectory

$$X_s = A(-t)^{\frac{1}{1-\theta}} = A(-t)^N, \tag{9.4.10}$$

where $N = \frac{1}{1-\theta}$ or $\theta = \frac{N-1}{N}$. As in the case of implosion, the time exponent N and the constant A cannot be determined from the energy integral as in the classical blast wave problem because the integral diverges. The value of θ or N is determined by seeking a solution that is regular at the singularity of the similarity equations. The constant A, however, has to be determined from matching the similarity solution with initial conditions. Since the initial motion is non-similar, it has to be determined numerically.

To facilitate the numerical integration of the similarity equations, the derivatives $\frac{d\psi}{d\xi}$, $\frac{d\phi}{d\xi}$, and $\frac{df}{d\xi}$ are first determined from Eqs 9.4.6–9.4.8. Doing so, we obtain

$$\frac{d\psi}{d\xi} = \frac{(\phi - \xi)[\theta\phi\psi - (\phi - \xi)\delta\psi] - [2\theta - (\gamma - 1)\delta]f}{(\phi - \xi)\left[(\phi - \xi)^2 - \frac{\gamma f}{\psi}\right]}, \tag{9.4.11}$$

$$\frac{d\phi}{d\xi} = \frac{(2\theta + \delta)\frac{f}{\psi} - (\phi - \xi)\theta\phi}{(\phi - \xi)^2 - \frac{\gamma f}{\psi}}, \tag{9.4.12}$$

$$\frac{df}{d\xi} = \frac{\gamma f\theta\phi - (\phi - \xi)(2\theta + \delta)f}{(\phi - \xi)^2 - \frac{\gamma f}{\psi}}. \tag{9.4.13}$$

The boundary conditions at the shock front $\xi = 1$ are given by the strong shock relationships

$$\psi(1) = \frac{\gamma + 1}{\gamma - 1}, \qquad \phi(1) = f(1) = \frac{2}{\gamma + 1}. \tag{9.4.14}$$

To integrate Eqs 9.4.11–9.4.13, a trial value of θ is first chosen and the equations are then integrated numerically in the range $1 \le \xi \le \infty$. The singularity $(\phi - \xi)^2 - \frac{\gamma f}{\psi} = 0$ will be encountered and the value of θ will be determined by

Table 9.3 Values of N

δ	$\gamma = 5/3$	$\gamma = 7/5$	$\gamma = 6/5$
3.25	0.590	—	—
2	0.696	0.718	0.752
1	0.816	0.831	0.855
0.5	0.817	0.906	0.920

iteration until the solution is regular at the singularity. At the singularity when the denominator of Eqs 9.4.11–9.4.13 vanish, we require the numerators to vanish simultaneously also, that is,

$$(\phi - \xi)[\theta\phi\psi - (\phi - \xi)\delta\psi] - [2\theta - (\gamma - 1)\delta]f \to 0,$$

$$(2\theta + \delta)\frac{f}{\psi} - (\phi - \xi)\theta\phi \to 0,$$

$$\gamma f\theta\phi - (\phi - \xi)(2\theta + \delta)f \to 0.$$

Using the condition $(\phi - \xi)^2 - \frac{\gamma f}{\psi} = 0$, it is found that all the numerators vanish if

$$\theta = \frac{\delta f}{(\phi - \xi)\phi\psi - 2f}. \tag{9.4.15}$$

The values of N for various values of γ and δ are given in Table 9.3.

As $X \to 0$, $\xi \to \infty$ and the limiting form of the solution for large values of ξ can be obtained from Eqs 9.4.11 to 9.4.13 as

$$\psi \sim \xi^{\delta},$$

$$\phi \sim \xi^{\theta},$$

$$f \sim \xi^{2\theta+\delta}.$$

In terms of the dimensional variables, the above equation can be expressed as

$$\rho = \rho_0\psi \sim x^{\delta},$$

$$u = \dot{X}_s\phi \sim -x^{\theta},$$

$$p = \rho_0\dot{X}_s^2 f \sim x^{2\theta+\delta}.$$

Thus as $X_s \to 0$, the density distribution is proportional to the original density profile $\rho_0 \sim x^{\delta}$. The particle velocity u tends to infinity as $x \to 0$ since $\theta < 0$. The pressure approaches zero as $x \to 0$ since in general $2\theta + \delta > 0$.

Similar to the implosion problem, the state variables Z, V can also be used, which reduces the integration to one equation for $Z(V)$. Once a solution for $Z(V)$ is obtained, the other variables ψ and ϕ can be determined by integrating the appropriate equations.

9.5 The Sharp Blow Problem

The solutions for both the implosion and the shock propagation into a non-uniform density medium problem are asymptotic self-similar solutions that are valid only in the region in the vicinity of the shock front. The existence of the singular characteristic divides the flow field preventing one region of the flow field from influencing the other.

Hence, a non-similar region of the flow field can be isolated from a self-similar region by the singular characteristic. An initially non-similar flow may become self similar asymptotically. Thus we say that the asymptotic self-similar solution is not uniformly valid. An excellent illustration of how an asymptotic self-similar solution becomes problematic when used to describe the entire flow field can be illustrated by the so-called "sharp blow problem" discussed in detail by Zel'dovich (1956).

Consider a gas occupying the positive half space $x > 0$ and a vacuum is in the negative half space $x < 0$. A finite energy per unit area, E_0, is deposited instantaneously at the interface $x = 0$ at $t = 0$ generating a strong shock wave that propagates to the right. Since the energy is released in a vanishingly small volume, the temperature (hence sound speed) is infinite at $x = 0$, and hence the escape front will propagate at infinite speed to the left for $t > 0$. The flow field is bounded by the shock and the escape front, that is, $-\infty \le x < x_s$. The problem is almost the same as the strong planar blast problem where a similarity solution exists. The similarity equations for planar motion are given by

$$(\phi - \xi)\psi' + \psi\phi' = 0, \tag{9.5.1}$$

$$(\phi - \xi)\phi' + \theta\phi + \frac{1}{\psi}f' = 0, \tag{9.5.2}$$

$$(\phi - \xi)f' + \gamma f\phi' + 2\theta f = 0, \tag{9.5.3}$$

where $\phi = \frac{u}{\dot{x}_s}$, $\psi = \frac{\rho}{\rho_0}$, $f = \frac{p}{\rho_0\dot{x}_s^2}$, and the primed quantities denote derivatives with respect to $\xi = \frac{x}{x_s}$. For a self-similar solution, $\theta = \frac{x_s\ddot{x}_s}{\dot{x}_s^2}$, and corresponds to a power law shock trajectory $x_s = At^N$ where $\theta = \frac{N-1}{N}$ or $N = \frac{1}{1-\theta}$, and $\theta =$ constant.

Solving for the derivatives ϕ', ψ', f' give

$$\psi' = \frac{-2\theta f + (\phi - \xi)\theta\phi\psi}{(\phi - \xi)\left[(\phi - \xi)^2 - \frac{\gamma f}{\psi}\right]}, \tag{9.5.4}$$

$$\phi' = \frac{-(\phi - \xi)\theta\phi + \frac{2\theta f}{\psi}}{(\phi - \xi)^2 - \frac{\gamma f}{\psi}}, \tag{9.5.5}$$

$$f' = \frac{-(\phi - \xi)2\theta f + \gamma\theta f\phi}{(\phi - \xi)^2 - \frac{\gamma f}{\psi}}. \tag{9.5.6}$$

Equations 9.5.4–9.5.6 can be integrated with the boundary condition at the shock front $\xi = 1$, that is, $\psi(1) = \frac{\gamma+1}{\gamma-1}$, $\phi(1) = f(1) = \frac{2}{\gamma+1}$, if the value of θ is known. Following the classical strong blast problem, we write the energy integral, as

$$E_0 = \int_{-\infty}^{x_s} \left(\frac{p}{\gamma - 1} + \frac{\rho u^2}{2}\right) dx = \rho_0\dot{x}_s^2 x_s \int_{-\infty}^{1} \left(\frac{f}{\gamma - 1} + \frac{\psi\phi^2}{2}\right) d\xi.$$

Table 9.4 Values of θ and N for Different Values of γ

γ	θ	N
1	−1.00	0.5
1.4	−0.667	0.6
1.666	−0.637	0.6107
3	−0.5949	0.627
7	−0.5723	0.636
∞	−0.5576	0.642

In the strong blast problem, the flow field is bounded by $0 \le \xi \le 1$, whereas in the present problem, $-\infty \le \xi \le 1$. If the solutions for $\psi(\xi)$, $\phi(\xi)$, and $f(\xi)$ are continuous and a finite value for the energy integral is obtained, then $\dot{x}_s^2 x_s$ must not be explicitly time dependent. Thus $\dot{x}_s^2 x_s \sim t^{3N-2}$ giving a value for $N = \frac{2}{3}, \theta = -\frac{1}{2}$ as in the classical strong planar blast solution. However, if this value of $\theta = -\frac{1}{2}$ is used in the integration of Eqs 9.5.4–9.5.6 with the strong shock condition at $\xi = 1$, the singularity $(\phi - \xi)^2 - \frac{\gamma f}{\psi} = 0$ will be encountered somewhere in the flow field $-\infty \le \xi \le 1$. Note that the singularity will not be encountered in the region $0 \le \xi \le 1$ permitting a continuous solution for the strong planar blast problem. The divergence obtained indicates in that the energy integral cannot be used to determine θ for the sharp blow problem. When the singularity is encountered using the value of $\theta = -\frac{1}{2}$ ψ', f', and ϕ' become infinite. Since there is no physical reason for the infinite gradients to be present, it is concluded that the value of $\theta = -\frac{1}{2}$ does not lead to an acceptable solution for the present problem. As in the implosion problem discussed previously, we must therefore seek a value of θ that gives a regular solution at the singularity.

When $(\phi - \xi)^2 - \frac{\gamma f}{\psi} \to 0$, we must require the numerators of Eqs 9.5.4–9.5.6 to vanish simultaneously, that is,

$$-2\theta f^* + (\phi^* - \xi^*)\theta\phi^*\psi^* \to 0, \tag{9.5.7}$$

$$(\phi^* - \xi^*)\theta\phi^* + \frac{2\theta f^*}{\psi^*} \to 0, \tag{9.5.8}$$

$$-2\theta f^* + (\phi^* - \xi^*) + \gamma f \phi^* \theta \to 0, \tag{9.5.9}$$

where we have used the asterisk, $*$, to denote the condition at the singularity, that is, $(\phi^* - \xi^*)^2 - \frac{\gamma f^*}{\psi^*} = 0$. Using the singularity condition, it can be shown that Eqs 9.5.7–9.5.9 can all be satisfied simultaneously if

$$\gamma\phi^* = 2(\phi^* - \xi^*). \tag{9.5.10}$$

To determine the desired solution, Eqs 9.5.4–9.5.6 are integrated numerically using the strong shock conditions at the front $\xi = 1$ with a trial value of θ. The desired solution will be the one that satisfies the regularity condition (i.e., Eq 9.5.10) at the singularity. The desired value of θ is obtained by iteration to the desired accuracy. The value of θ (or N) for various values of γ are given in Table 9.4. Note that $\frac{1}{2} \le N \le \frac{2}{3}$, and for $N = \frac{2}{3}$

the solution corresponds to the classical constant energy planar blast solution. Thus the solution that is regular at the singularity will not satisfy the energy integral.

9.6 Exact Solution for $\gamma = 1.4$

For the particular value of $\gamma = 1.4$, von Hoerner (1955) found an analytical solution for the sharp blow problem. The velocity profile is linear and can be expressed in the following form

$$\phi = a\xi + b. \tag{9.6.1}$$

The solution for ψ and f can be obtained by substituting Eq 9.6.1 into the continuity equation (Eq 9.5.1) which integrates to yield

$$\psi = K\left[(a-1)\xi + b\right]^{\frac{-a}{a-1}}. \tag{9.6.2}$$

Substituting the velocity and density profiles into the momentum equation (Eq 9.5.2) and integrating gives

$$f = K_1\left[(a-1)\xi + b\right]^{-\left(\frac{\gamma a+2\theta}{a-1}\right)}. \tag{9.6.3}$$

The integration constants a, b, K, and K_1 can be evaluated using the boundary condition at the shock. For $\gamma = 1.4$, we obtained

$$\phi(\xi) = \frac{5}{6}(2\xi - 1), \tag{9.6.4}$$

$$\psi(\xi) = 6(5 - 4\xi)^{-5/2}, \tag{9.6.5}$$

$$f(\xi) = \frac{5}{6}(5 - 4\xi)^{-3/2}. \tag{9.6.6}$$

The exact solution for $\gamma = 1.4$ is given by Eqs 9.6.4–9.6.6 and is shown in Fig. 9.4. The location of the singularity ξ^* can now be obtained by substituting the solution of Eqs 9.6.4–9.6.6 into the singularity equation, that is,

$$\xi^* = \phi^* \pm \left(\frac{\gamma f^*}{\psi^*}\right)^{\frac{1}{2}}.$$

Doing so, we obtain $\xi^* = 5/4$ or $-1/2$. And since the flow is bounded by $-\infty \leq \xi \leq 1$, we may discard the value of $\xi^* = 5/4$. Substituting $\xi^* = -1/2$ and $\phi^* = 5/6(2\xi^* - 1) = -5/3$ into Eq 9.5.10 shows that the regularity condition is satisfied.

Since the solution for $\phi(\xi)$, $\psi(\xi)$, and $f(\xi)$ are given in simple analytical forms by Eqs 9.6.4–9.6.6, we may investigate the various integrals of the motion. Since the flow follows the shock to the right and is also accelerated to the left by the escape front, there will be a boundary where the particle velocity is zero. Solving for $\xi = \xi_0$ where $\phi = 0$ from Eq 9.6.4, we obtain $\xi_0 = 1/2$. Thus, the mass between $1/2 \leq \xi \leq 1$ moves to the

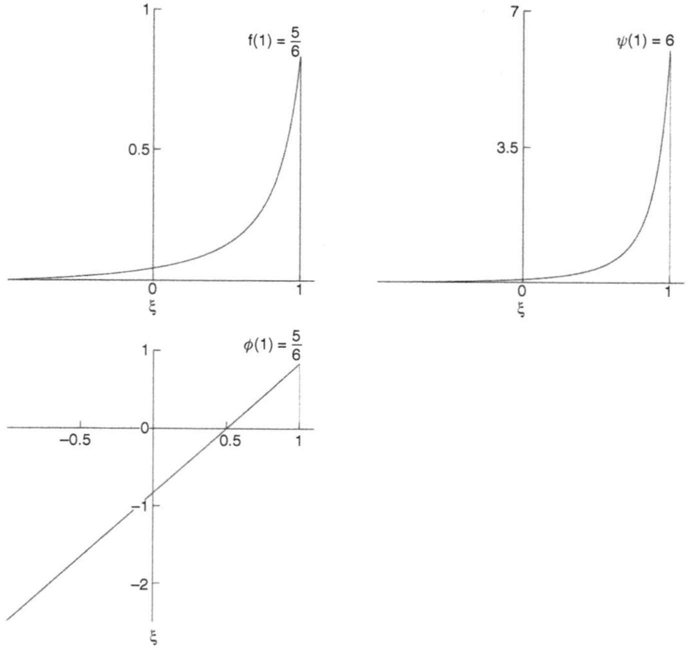

Figure 9.4 The Exact Solution for $\gamma = 1.4$

left with the shock and the mass between $-\infty \le \xi \le 1/2$ moving to the left. The mass moving to the right is given by the integral

$$M_s = \int_{x_0}^{x_s} \rho \mathrm{d}x = \rho_0 x_s \int_{\xi_0=1/2}^{\xi=1} \psi \mathrm{d}\xi,$$

and with $\psi(\xi)$ given by Eq 9.6.5, we can evaluate the integral and obtain

$$M_s = \rho_0 x_s 0.8075 = A\rho_0 0.8075 t^{0.6}, \tag{9.6.7}$$

where we have used $x_s = At^N$ and $N = 0.6$ for this particular case of $\gamma = 1.4$. Equation 9.6.7 indicates that M_s increases with time and since the velocity ahead of the shock is zero, and also at the boundary, $\xi = 1/2$, the momentum of the fluid between $1/2 \le \xi \le 1$ will increase with time. The momentum integral can be written as

$$I = \int_{x_0}^{x_s} \rho u \mathrm{d}x = \rho_0 \dot{x}_s x_s \int_{\xi_0=0.5}^{\xi=1} \psi\phi \mathrm{d}\xi.$$

With $\psi(\xi)$ and $\phi(\xi)$ given by Eqs 9.6.5 and 9.6.6, the integral can be evaluated. Denoting the dimensionless integral

$$\int_{0.5}^{1} \psi\phi \mathrm{d}\xi = B,$$

and noting that $x_s = At^N$, $\dot{x}_s = ANt^{N-1}$, the momentum integral can be written as

$$I_0 = (\rho_0 A^2 BN)t^{2N-1}.$$

For the momentum to increase with increasing time, $N > 1/2$ and indeed, the values of N for $1 \le \gamma \le \infty$ shown in Table 9.4 indicate that $N > 1/2$.

Similarly, we may evaluate the energy in the volume between $0.5 \le \xi \le 1$, that is,

$$E = \rho_0 \dot{x}_s^2 x_s \int_{0.5}^{1} \left(\frac{\psi \phi^2}{2} + \frac{f}{\gamma - 1} \right) d\xi = (\rho_0 A^3 N^2 C)t^{3N-2},$$

where C is the dimensionless integral

$$C = \int_{0.5}^{1} \left(\frac{\psi \phi^2}{2} + \frac{f}{\gamma - 1} \right) d\xi.$$

Since there is no continuous energy input to the flow after the initial instant, the energy in the volume $0.5 \le \xi \le 1$ must decrease with time as the shock decays. For E to decrease with increasing time, the exponent $3N - 2 < 0$, that is, $N < 2/3$. Again the values of N shown in Table 9.4 satisfy this requirement. Thus we see that $1/2 \le N \le 2/3$ and is in accord with the values given in Table 9.4.

9.7 Determination of A

The asymptotic self-similar solution is determined from the basic conservation equations and the value of the constant A in the shock trajectory (i.e., $x_s = At^N$) is not known. The value of A has to be determined from a consideration of the initial conditions of the problem. In the classical strong blast solution, the value of A is obtained from the energy integral. For the present problem, the energy integral is written as

$$E_0 = \int_{-\infty}^{x_s} \left(\frac{p}{\gamma - 1} + \frac{\rho u^2}{2} \right) dx = \rho_0 \dot{x}_s^2 x_s \int_{-\infty}^{1} \left(\frac{f}{\gamma - 1} + \frac{\psi \phi^2}{2} \right) d\xi.$$

If the dimensionless integral is constant and finite, then dimensional considerations give $N = 2/3$ and the value of A equal to

$$A = \left(\frac{E_0}{\rho_0 N^2 B} \right)^{\frac{1}{3}},$$

where

$$B = \int_{-\infty}^{1} \left(\frac{f}{\gamma - 1} + \frac{\psi \phi^2}{2} \right) d\xi.$$

However the integral, B, diverges for the sharp blow problem because the kinetic energy tends to infinity near the vacuum boundary faster than the density goes to zero. This can be demonstrated by seeking the solution near the vacuum interface.

Near the escape front, the pressure gradient is vanishingly small and neglecting it, the momentum equation (i.e., Eq 9.5.2) becomes

$$(\phi - \xi)\phi' + \theta\phi = 0,$$

which can be integrated to yield

$$\phi = (1 - \theta)\xi = \frac{\xi}{N}. \tag{9.7.1}$$

Substituting the above solution for ϕ into the continuity equation (i.e., Eq 9.5.1) gives

$$\frac{d\psi}{\psi} = \frac{\theta}{1-\theta}\frac{d\xi}{\xi},$$

which integrates to give the density profile near the escape front, as

$$\psi = K|\xi|^{\frac{1-\theta}{\theta}} = K|\xi|^{\frac{-1}{1-N}}. \tag{9.7.2}$$

In the above equation, K is a constant of integration and we write the absolute value for ξ to ensure a positive density.

With the solution for ϕ and ψ, the pressure profile can be obtained from the momentum equation as

$$f = K_1|\xi|^{2-\gamma\left(\frac{1-\theta}{\theta}\right)} = K_1|\xi|^{2-\frac{\gamma}{1-N}}, \tag{9.7.3}$$

where K_1 is another integration constant. The above asymptotic solution near the escape front is valid for arbitrary values of γ.

For the particular value of $\gamma = 1.4$ where closed formed solutions can be obtained, Eqs 9.6.4–9.6.6 reduces to the above asymptotic solution (i.e., Eqs 9.7.1–9.7.3) for large values of $|\xi|$. From Eqs 9.7.1 and 9.7.2, we see that the kinetic energy near the escape front $\frac{\phi^2\psi}{2} \sim \xi^{2-\frac{1}{1-N}} \sim \xi^{\frac{1-2N}{1-N}}$ and for the values of $1/2 \leq N \leq 2/3$, the kinetic energy approaches infinity as $\xi \to \infty$. Thus the energy integral diverges even though the solutions for ϕ, ψ, f are continuous and singularity free. Hence, unlike the classical strong blast problem, the energy integral cannot be used now to evaluate A in the sharp blow problem.

The infinite energy result suggests that the self-similar solution is not uniformly valid in the entire range $-\infty \leq \xi \leq 1$. Indeed, the flow near the vacuum edge must retain the memory of the initial energy release process. The entropy of the fluid near the escape front is due to the initial energy deposition process and differs from the entropy of the fluid elsewhere in the flow field which is due to the shock compression process. To elucidate the initial non-similar energy release process, let us assume that the initial energy release is due to the impulsive motion of a piston acting on the gas for a short duration, τ. The impulse and the energy imparted to the gas by the piston can be written as

$$I = \int_0^\tau p_p \mathrm{d}t, \tag{9.7.4}$$

$$E = \int_0^\tau p_p u_p \mathrm{d}t, \tag{9.7.5}$$

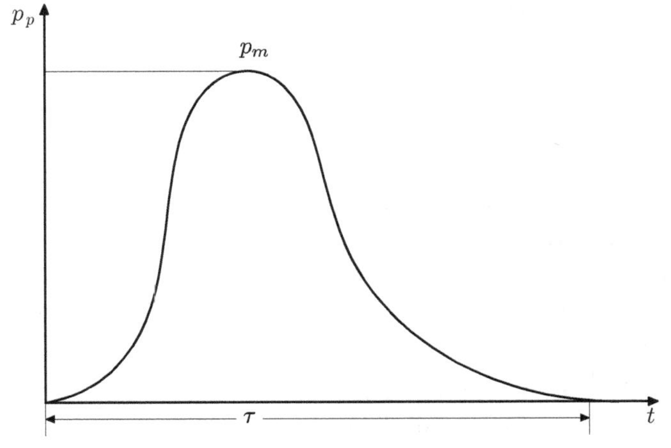

Figure 9.5 Pressure–Time Profile

where p_p and u_p denote the pressure at the piston face and the piston velocity, respectively. Given the pressure–time profile, as shown in Fig. 9.5, the impulse and the energy can be evaluated.

We may write Eqs 9.7.4 and 9.7.5 as

$$I = p_m \tau \int_0^1 \bar{p}_p \mathrm{d}\bar{t} = a p_m \tau, \tag{9.7.6}$$

$$E = \frac{p_m^{3/2}}{\rho_0^{1/2}} \tau \int \bar{p}_p \bar{u}_p \mathrm{d}\bar{t} = b \frac{p_m^{3/2}}{\rho_0^{1/2}} \tau, \tag{9.7.7}$$

where $\bar{p}_p = \frac{p_p}{p_m}$ and $\bar{t} = \frac{t}{\tau}$. We normalized the piston velocity using the particle velocity behind the shock $u_1 = \frac{2}{\gamma+1}\dot{x}_s$. Since the shock pressure $p_s = \frac{2}{\gamma+1}\rho_0 \dot{x}_s^2$ and $p_s \simeq p_p \simeq p_m$, $u_1 \simeq (\frac{p_m}{\rho_0})^{\frac{1}{2}}$. The dimensionless integrals a and b depend on the nature of the pressure–time profile of the piston. For example, for a simple square pressure pulse, shown in Fig. 9.6, that is, $a = 1$ and $b = (\frac{2}{\gamma+1})^{\frac{1}{2}}$. In general, a and b are of the order of

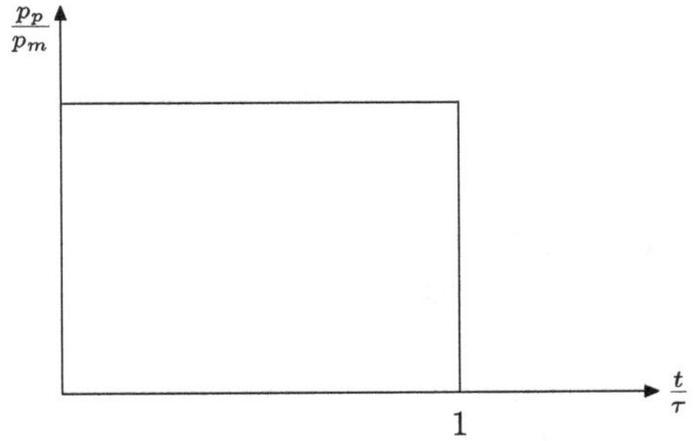

Figure 9.6 Simple Square Pulse

unity and we may write Eqs 9.7.6 and 9.7.7 as

$$I \simeq p_m \tau,$$

$$E \sim \frac{p_m^{3/2}}{\rho_0^{1/2}} \tau.$$

In the self-similar solution, the shock trajectory is given by $\dot{x}_s = At^N$ and thus the shock velocity $\dot{x}_s = ANt^{N-1}$. With $\dot{x}_s \sim (\frac{p_m}{\rho_0})^{\frac{1}{2}}$ hence $ANt^{N-1} \sim (\frac{p_m}{\rho_0})^{\frac{1}{2}}$, $A \sim (\frac{p_m}{\rho_0})^{\frac{1}{2}} t^{1-N}$. If $A =$ constant as $t \to 0$, then p_m must approach infinity like $t^{-2(1-N)}$ as $t \to 0$ in this case. For $N > \frac{1}{2}$, $I = p_m \tau \to \tau^{2N-1}$ as $t \to 0$ in this case. In other words, the impulse $I \to 0$ as $t \to 0$ in this case. Since the energy $E \sim \frac{p_m^{3/2}}{\rho_0^{1/2}} \tau \sim t^{3N-2}$ as $t \to 0$, $E \to \infty$ as $t \to 0$, for $N < 2/3$, that is, the energy imparted to the gas by the piston is infinite. Thus if the self-similar solution exists in the limit as $\frac{t}{\tau} \to 0$, it is evident that the self-similar solution cannot be uniformly valid in space encompassing all the mass set in motion from the shock front to the vacuum interface. To resolve this dilemma, Zeldovich argued that the self-similar solution should only include the mass described by the self-similar motion. A small portion of the mass near the escape front that bears the "imprint" of the initial non-similar energy deposition process should be neglected. The amount of mass to be neglected is determined by requiring that the energy integral is time independent as the self-similar solution demands. The limits of the integration is then taken to be $\xi_0 \leq \xi \leq 1$ instead of $-\infty \leq \xi \leq 1$. The mass to be neglected is between $-\infty \leq \xi \leq \xi_0$. By restricting the limits of the integration, the integral can now be made finite. The proper procedure to relate the self-similar solution to the non-similar initial motion is to solve the non-similar motion numerically and match it with the self-similar solution.

9.8 Converging Blast Waves

The similarity solution described previously in Section 9.2 is valid only near the vicinity of the center of convergence as $R_s \to 0$. Furthermore, the solution can describe only a limited region of the flow field behind the shock. To describe the initial stages of the propagation of a converging shock wave, a different solution has to be constructed that considers the details of how the shock wave is generated, for example, impulsive motion of a converging piston or the instantaneous deposition of energy on the inner surface of a spherical (or cylindrical shell). We shall investigate the propagation of a converging blast wave in this section.

Assume at time $t = 0$, an energy per unit area, E_0, is released instantaneously at the inner surface of a spherical (or cylindrical) cavity of radius R_0. A strong blast wave is generated which decays as it propagates inwards away from the surface at later times. For very early times when $R_s \simeq R_0$, the blast wave is essentially a strong planar blast wave. Curvature and finite Mach number effects come in progressively as the blast propagates away from the wall at later times. This is illustrated in Fig. 9.7. The very early time shock propagation can be described by the self-similar solution for a strong

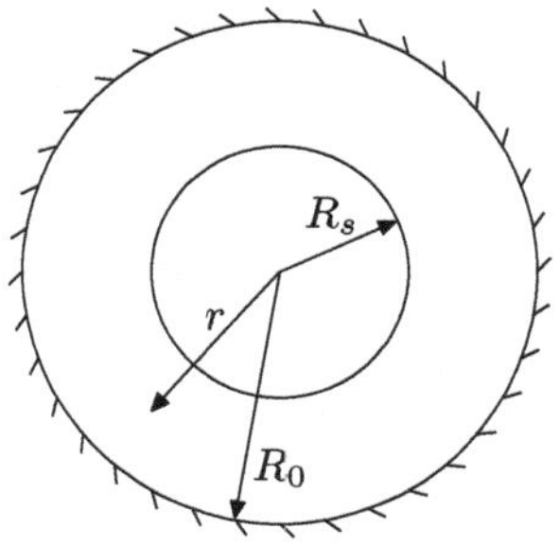

Figure 9.7 Curvature Effects as Blast Propagates Away from Initial Position

planar blast given in Chapter 4. This similarity solution can be perturbed to account for curvature effects at later times.

For the initial blast motion, that is, $R_s \simeq R_0$, the basic equations are given by

$$\frac{\partial \rho}{\partial t} + \frac{\partial(\rho u)}{\partial r} + \frac{j\rho u}{r} = 0,$$

$$\frac{\partial u}{\partial t} + u\frac{\partial u}{\partial r} + \frac{1}{\rho}\frac{\partial p}{\partial r} = 0,$$

$$\left(\frac{\partial}{\partial t} + u\frac{\partial}{\partial r}\right)\frac{p}{\rho^{\gamma}} = 0,$$

where $j = 1, 2$ for cylindrical and spherical geometry, respectively. Defining the following blast wave variables

$$\left.\begin{aligned} f(\xi, x_s) &= \frac{p(r,t)}{\rho_0 \dot{x}_s^2} \\ \xi &= \frac{r - R_0}{R_s - R_0} \\ X_s &= R_s - R_0 \\ x_s &= \frac{X_s}{R_0} = \frac{R_s - R_0}{R_0} \end{aligned}\right\} \tag{9.8.1}$$

the conservation equations become

$$(\phi - \xi)\frac{\partial \psi}{\partial \xi} + \psi\frac{\partial \phi}{\partial \xi} + \frac{j\phi\psi x_s}{1 + x_s\xi} = -x_s\frac{\partial \psi}{\partial x_s}, \tag{9.8.2}$$

$$(\phi - \xi)\frac{\partial \phi}{\partial \xi} + \theta\phi + \frac{1}{\psi}\frac{\partial f}{\partial \xi} = -x_s\frac{\partial \phi}{\partial x_s}, \tag{9.8.3}$$

$$(\phi - \xi)\left(\frac{\partial f}{\partial \xi} - \frac{\gamma f}{\psi}\frac{\partial \psi}{\partial \xi}\right) + 2\theta f = -x_s\left(\frac{\partial f}{\partial x_s} - \frac{\gamma f}{\psi}\frac{\partial \psi}{\partial x_s}\right), \tag{9.8.4}$$

where $\theta = \frac{x_s\ddot{x}_s}{\dot{x}_s^2}$. A more convenient form of the energy equation can be obtained by eliminating ϕ from Eq 9.8.4 using the continuity equation, that is,

$$(\phi - \xi)\frac{\mathrm{d}f}{\mathrm{d}\xi} + \gamma f \frac{\mathrm{d}\phi}{\mathrm{d}\xi} + 2\theta f + \frac{j\gamma f \phi x_s}{1 + x_s\xi} = -x_s \frac{\mathrm{d}f}{\mathrm{d}x_s}. \tag{9.8.5}$$

Note that Eqs 9.8.2–9.8.4 (or 9.8.5) are still partial differential equations, the same as the regular Euler equations in dimensional parameters. The flow field is bounded by the shock front and the wall, that is, $0 \leq \xi \leq 1$. At the shock front $\xi = 1$, the boundary conditions are given by the Rankine–Hugoniot relationships, that is,

$$\frac{\rho_1}{\rho_0} = \psi(1, x_s) = \frac{\gamma + 1}{(\gamma - 1) + 2/M_s^2}, \tag{9.8.6}$$

$$\frac{u_1}{\dot{x}_s} = \phi(1, x_s) = \frac{2}{\gamma + 1}\left(1 - \frac{1}{M_s^2}\right), \tag{9.8.7}$$

$$\frac{p_1}{\rho_0 \dot{x}_s^2} = f(1, x_s) = \frac{2}{\gamma + 1} - \frac{\gamma - 1}{\gamma(\gamma + 1)} \frac{1}{M_s^2}, \tag{9.8.8}$$

where M_s is the shock Mach number defined as

$$M_s = \frac{|\dot{X}_s|}{c_0} = \frac{|\dot{X}_s|}{\left(\frac{\gamma p_0}{\rho_0}\right)^{\frac{1}{2}}}.$$

The energy integral is written as

$$E_0 = \int_{R_s}^{R_0} \left(\frac{p}{\gamma - 1} + \frac{\rho u^2}{2}\right) k_j \gamma^j \mathrm{d}r - \int_{R_s}^{R_0} \frac{p_0}{\gamma - 1} k_j r^j \mathrm{d}r,$$

and in terms of the dimensionless parameters ϕ, f, ψ, ξ, the energy integral becomes

$$E_0 = \rho_0 k_j R_0^{j+3} \dot{x}_s^2 x_s \left[I(x_s) - \frac{c_0^2 \left(1 - (1 + x_s)^{j+1}\right)}{R_0^2 \gamma(\gamma - 1)(j + 1)\dot{x}_s^2 x_s} \right], \tag{9.8.9}$$

where

$$I(x_s) = \int_1^0 \left(\frac{f}{\gamma - 1} + \frac{\psi\phi^2}{2}\right)(1 + x_s\xi)^j \,\mathrm{d}\xi.$$

In the above equation $k_j = 2\pi,\ 4\pi$ for cylindrical and spherical geometry respectively. For strong shocks where $\frac{c_0^2}{\dot{x}_s^2} \ll 1$, the second term inside the bracket on the right hand side of the integral can be neglected. However, the integral $I(x_s)$ is still time (or x_s) dependent due to the presence of the characteristic length scale R_0. When $x_s \lll 1$, then $I(x_s) \to$ constant and a self-similar solution is possible. Thus, departure from similarity arises from both: the presence of a characteristic length scale R_0 and for finite shock strength.

It is more convenient to define a characteristic time t_j^* as

$$t_j^* = \left(\frac{\rho_0 k_j R_0^{j+3}}{(j+1)E_0} \right)^{\frac{1}{2}}. \tag{9.8.10}$$

Note that the term within brackets in Eq 9.8.10 is just the inverse of the energy density, that is, $\frac{E_0}{\rho_0 k_j R_0^{j+1}/(j+1)}$ and t^* can be interpreted as the implosion time. Non-dimensionalizing time t by t_j^*, that is, $\tau = \frac{t}{t_j^*}$, we note the shock velocity $\dot{R}_s = \dot{x}_s = R_0 \dot{x}_s = \frac{R_0}{t_j^*} \frac{\mathrm{d}x_s}{\mathrm{d}\tau}$.

The conservation equations remain the same except $\dot{x}_s$ is now $\frac{\mathrm{d}x_s}{\mathrm{d}\tau}$ instead of $\frac{\mathrm{d}x_s}{\mathrm{d}t}$. The energy integral can conveniently be expressed now as

$$1 = \dot{x}_s^2 x_s (j+1) I(x_s) - \frac{c_j^{*2} \left(1 - (1+x_s)^{j+1}\right)}{\gamma(\gamma-1)}, \tag{9.8.11}$$

where $c_j^{*2} = \frac{c_0^2 t_j^{*2}}{R_0^2}$. For large E_0, $t_j^* \to 0$, $c_j^* \to 0$, and the second term on the right hand side of Eq 9.8.11 vanishes. However, a self-similar solution is still not possible because $I(x_s)$ is a function of x_s.

For small times where $x_s \ll 1$, we may seek solutions as a power series in x_s, that is,

$$\left.\begin{aligned} \phi(\xi, x_s) &= \sum \phi^{(n)}(\xi) x_s^n \\ f(\xi, x_s) &= \sum f^{(n)}(\xi) x_s^n \\ \psi(\xi, x_s) &= \sum \psi^{(n)}(\xi) x_s^n \end{aligned}\right\} \tag{9.8.12}$$

For $x_s \to 0$, $I(x_s) \to$ constant and the energy integral (Eq 9.8.11) indicates that $\dot{x}_s^2 x_s \to$ constant as $x_s \to 0$ or $\dot{x}_s^2 \to 1/x_s$ as $x_s \to 0$. Thus the form for the expansion of $\dot{x}_s^2$ is

$$\dot{x}_s^2 = \frac{1}{x_s} \Sigma F_n x_s^n = \frac{F_0}{x_s} + F_1 + F_2 x_s + \cdots \tag{9.8.13}$$

Accordingly $\theta = \frac{x_s \ddot{x}_s}{\dot{x}_s^2}$ becomes

$$\theta = \Sigma \theta(n) x_s^n = \theta^{(0)} + \theta^{(1)} x_s + \theta^{(2)} x_s^2 + \theta^{(3)} x_s^3 + \cdots \tag{9.8.14}$$

where

$$\theta^{(0)} = -\frac{1}{2}, \qquad \theta^{(1)} = \frac{F_1}{2F_0}, \qquad \theta^{(2)} = \frac{F_2}{F_0} - \frac{F_1^2}{2F_0^2},$$

$$\theta^{(3)} = \frac{1}{2}\left[\left(\frac{F_1}{F_0}\right)^3 - 3\frac{F_1 F_2}{F_0^2} + 3\frac{F_3}{F_0} \right]. \tag{9.8.15}$$

Substituting the various perturbation expressions into the conservation equations and the energy integral, we obtain after equating coefficients of similar order in x_s the following:

Zeroth order in x_s

$$\left.\begin{aligned}
&\left(\phi^{(0)}-\xi\right)\psi'^{(0)}+\psi^{(0)}\phi'^{(0)}=0\\
&\left(\phi^{(0)}-\xi\right)\phi'^{(0)}+\frac{1}{\psi^{(0)}}f'^{(0)}=-\theta^{(0)}\phi^{(0)}\\
&\left(\phi^{(0)}-\xi\right)f'^{(0)}+\gamma f^{(0)}\phi'^{(0)}=-2\theta^{(0)}f^{(0)}\\
&F_0=\frac{1}{(j+1)}I^{(0)}\\
&I^{(0)}=\int_1^0\left(\frac{f^{(0)}}{\gamma-1}+\frac{\psi^{(0)}\phi^{(0)}}{2}\right)^2\mathrm{d}\xi
\end{aligned}\right\}\tag{9.8.16}$$

First order in x_s

$$\left.\begin{aligned}
&\left(\phi^{(0)}-\xi\right)\psi'^{(1)}+\psi^{(0)}\phi'^{(1)}\\
&\quad=-\left[\psi^{(1)}\left(\phi'^{(0)}+1\right)+\phi^{(1)}\psi'^{(0)}+j\phi^{(0)}\psi^{(0)}\right]\\
&\left(\phi^{(0)}-\xi\right)\phi'^{(1)}+\frac{1}{\psi^{(0)}}f'^{(1)}\\
&\quad=-\left[\phi^{(1)}\left(\phi'^{(0)}+\theta^{(0)}\right)+\theta^{(1)}\phi^{(0)}-\frac{\psi^{(1)}f'^{(0)}}{\psi^{(0)2}}\right]\\
&\left(\phi^{(0)}-\xi\right)f'^{(1)}+\gamma f^{(0)}\phi'^{(1)}=-\left[f^{(1)}\left(\gamma\phi'^{(0)}+1+2\theta^{(0)}\right)\right.\\
&\quad\left.+2\theta^{(1)}f^{(0)}+\phi^{(1)}f'^{(0)}+\gamma jf^{(0)}\theta^{(0)}\right]\\
&F_1=-\frac{1}{I^{(0)}}\left(\frac{c_j^{*2}}{\gamma(\gamma-1)}+F_0I^{(1)}\right)\\
&I^{(1)}=\int_1^0\left[j\xi\left(\frac{f^{(0)}}{\gamma-1}+\frac{\phi^{(0)2}\psi^{(0)}}{2}\right)\right.\\
&\quad\left.+\frac{f^{(1)}}{\gamma-1}+\frac{\phi^{(0)2}\psi^{(0)}}{2}+\phi^{(0)}\phi^{(1)}\psi^{(0)}\right]\mathrm{d}\xi
\end{aligned}\right\}\tag{9.8.17}$$

<u>Second order in x_s</u>

$$
\left.
\begin{aligned}
&\left(\phi^{(0)}-\xi\right)\psi'^{(2)}+\psi^{(0)}\phi'^{(2)}\\
&\quad=-\Big\{\psi^{(2)}\left(\phi'^{(0)}+2\right)+\psi^{(1)}\phi'^{(1)}+\phi^{(2)}\psi'^{(0)}\\
&\qquad+\phi^{(1)}\psi'^{(1)}-j\xi\phi^{(0)}\psi^{(0)}+j\left(\phi^{(0)}\psi^{(1)}+\phi^{(1)}\psi^{(0)}\right)\Big\}\\[2ex]
&\left(\phi^{(0)}-\xi\right)\phi'^{(2)}+\frac{f'^{(2)}}{\psi^{(0)}}\\
&\quad=-\Bigg\{\phi^{(2)}\left(\phi'^{(0)}+\theta^{(0)}2\right)+\phi^{(1)}\left(\phi'^{(1)}+\theta^{(1)}\right)\\
&\qquad+\phi^{(0)}\theta^{(2)}-\frac{\psi^{(1)}f'^{(1)}}{\psi^{(0)2}}+\left(\frac{\psi'^{(1)2}}{\psi^{(0)}}-\frac{\psi^{(2)}f'^{(0)}}{\psi^{(0)2}}\right)\Bigg\}\\[2ex]
&\left(\phi^{(0)}-\xi\right)f'^{(2)}+\gamma f^{(0)}\phi'^{(2)}\\
&\quad=-\Big\{f^{(2)}\left(\gamma\phi'^{(0)}+2\theta^{(0)}+2\right)\\
&\qquad+f^{(1)}\left(\gamma\phi'^{(1)}+2\theta^{(1)}+2\theta^{(2)}f^{(0)}\right)+\phi^{(2)}f'^{(0)}+\phi^{(1)}f'^{(1)}\\
&\qquad-\gamma j\xi\phi^{(0)}f^{(0)}+\gamma j\left(\phi^{(0)}f^{(1)}+\phi^{(1)}f^{(0)}\right)\Big\}\\[2ex]
&F_2=-\frac{1}{I^{(0)}}\left[\frac{jc_j^{*2}}{2\gamma(\gamma-1)}+F_1I^{(1)}+F_0I^{(2)}\right]\\[2ex]
&I_2^{(2)}=\int_1^0\mathrm{d}\xi\Bigg[\frac{f^{(2)}}{\gamma-1}+\frac{1}{2}\psi^{(2)}\phi^{(0)2}+\frac{1}{2}\psi^{(0)}\phi^{(1)2}\\
&\qquad+\psi^{(0)}\phi^{(0)}\phi^{(2)}+\psi^{(1)}\phi^{(1)}\phi^{(0)}\\
&\qquad+j\xi\left\{\frac{f^{(1)}}{\gamma-1}+\frac{1}{2}\psi^{(1)}\phi^{(0)2}+\psi^{(0)}\phi^{(0)}\phi^{(1)}\right\}\\
&\qquad+\frac{1}{2}j(j-1)\xi^2\left\{\frac{f^{(0)}}{\gamma-1}+\frac{1}{2}\psi^{(0)}\phi^{(0)2}\right\}\Bigg]
\end{aligned}
\right\}
\qquad (9.8.18)
$$

Third order in x_s

$$
\left.\begin{aligned}
&\left(\phi^{(0)}-\xi\right)\psi'^{(3)}+\psi^{(0)}\phi'^{(3)}=-\Big\{\psi^{(3)}\left(\phi'^{(0)}+3\right)+\psi^{(2)}\phi'^{(1)}+\psi^{(1)}\phi'^{(2)}\\
&\quad+\phi^{(3)}\psi'^{(0)}+\phi^{(2)}\psi'^{(1)}+\phi^{(1)}\psi'^{(2)}+j\left(\phi^{(0)}\psi^{(2)}+\phi^{(1)}\psi^{(1)}+\phi^{(2)}\psi^{(0)}\right)\\
&\quad-j\xi\left(\phi^{(0)}\psi^{(1)}+\phi^{(1)}\psi^{(0)}\right)+j\xi^2\phi^{(0)}\psi^{(0)}\Big\}\\
&\left(\phi^{(0)}-\xi\right)\phi'^{(3)}+\frac{f'^{(3)}}{\psi^{(0)}}=-\Big\{\phi^{(3)}\left(\phi'^{(0)}+3\right)+\psi^{(2)}\phi'^{(1)}+\psi^{(1)}\phi'^{(2)}\\
&\quad+\phi^{(3)}\psi'^{(0)}+\phi^{(2)}\psi'^{(1)}+\phi^{(1)}\psi'^{(2)}+j\left(\phi^{(0)}\psi^{(2)}+\phi^{(1)}\psi^{(1)}+\phi^{(2)}\psi^{(0)}\right)\\
&\quad-j\xi\left(\phi^{(0)}\psi^{(1)}+\phi^{(1)}\psi^{(0)}\right)+j\xi^2\phi^{(0)}\psi^{(0)}\Big\}\\
&\left(\phi^{(0)}-\xi\right)f'^{(3)}+\gamma f^{(0)}\phi'^{(3)}=-\Big\{f^{(3)}\left(\gamma\phi'^{(0)}+2\theta^{(0)}+3\right)\\
&\quad+f^{(2)}\left(\gamma\phi'^{(1)}+2\theta^{(1)}\right)+f^{(1)}\left(\gamma\phi'^{(2)}+2\theta^{(2)}\right)+2\theta^{(3)}f^{(0)}+\phi^{(3)}f'^{(0)}\\
&\quad+\phi^{(2)}f'^{(2)}+\gamma j\xi^2\phi^{(0)}f^{(0)}-\gamma j\xi\left(\phi^{(0)}f^{(1)}+\phi^{(1)}f^{(0)}\right)\\
&\quad+\gamma j\xi\left(\phi^{(0)}f^{(2)}+\phi^{(1)}f^{(1)}+\phi^{(2)}f^{(0)}\right)\Big\}\\
&F_3=-\left\{\frac{j(j-1)c_j^{*2}}{6\gamma(\gamma-1)}+F_0I^{(3)}+F_1I^{(2)}+F_2I^{(1)}\right\}\frac{1}{I^{(0)}}\\
&I^{(3)}=\int_1^0\mathrm{d}\xi\left[\frac{f^{(3)}}{\gamma-1}+\frac{1}{2}\psi^{(3)}\phi^{(0)2}+\frac{1}{2}\psi^{(1)}\phi^{(1)2}+\psi^{(0)}\phi^{(0)}\phi^{(3)}\right.\\
&\quad+\psi^{(0)}\phi^{(1)}\phi^{(2)}+\psi^{(1)}\phi^{(0)}\phi^{(2)}+\psi^{(2)}\phi^{(0)}\phi^{(0)}\\
&\quad+j\xi\left\{\frac{f^{(2)}}{\gamma-1}+\frac{1}{2}\psi^{(2)}\phi^{(0)2}+\frac{1}{2}\psi^{(0)}\phi^{(1)2}+\psi^{(0)}\phi^{(0)}\phi^{(2)}+\psi^{(1)}\phi^{(1)}\phi^{(0)}\right\}\\
&\quad\left.+\frac{1}{2}j(j-1)\xi^2\left\{\frac{f^{(1)}}{\gamma-1}+\frac{1}{2}\psi^{(1)}\phi^{(0)2}+\psi^{(0)}\phi^{(0)}\phi^{(1)}\right\}\right]
\end{aligned}\right\}
\tag{9.8.19}
$$

The boundary conditions at the shock front, $\xi = 1$, for the various orders can be obtained by expanding the Rankine–Hugoniot relationships (Eqs 9.8.8–9.8.10). Substituting Eq 9.8.13 into Eqs 9.8.8–9.8.10 and expanding in x_s, we obtain for

Zeroth order in x_s

$$\left.\begin{aligned} \phi^{(0)}(1) = f^{(0)}(1) &= \frac{2}{\gamma+1} \\ \psi^{(0)}(1) &= \frac{\gamma+1}{\gamma-1} \end{aligned}\right\} \tag{9.8.20}$$

First order in x_s

$$\left.\begin{aligned} \phi^{(1)}(1) &= \frac{-2c_j^{*2}}{F_0(\gamma+1)} \\ \psi^{(1)}(1) &= \frac{-2c_j^{*2}(\gamma+1)}{F_0(\gamma-1)^2} \\ f^{(1)}(1) &= \frac{1}{2}\frac{(\gamma-1)\phi^{(1)}(1)}{\gamma} \end{aligned}\right\} \tag{9.8.21}$$

Second order in x_s

$$\left.\begin{aligned} \psi^{(2)}(1) &= \frac{2c_j^{*2}(\gamma+1)}{(\gamma-1)^2F_0^2}\left(F_1 + \frac{2c_j^{*2}}{\gamma-1}\right) \\ \phi^{(2)}(1) &= \frac{2c_j^{*2}F_1}{(\gamma+1)F_0^2} \\ f^{(2)}(1) &= \frac{1}{2}\frac{(\gamma-1)\phi^{(2)}(1)}{\gamma} \end{aligned}\right\} \tag{9.8.22}$$

Third order in x_s

$$\left.\begin{aligned} \psi^{(3)}(1) &= \frac{-2c_j^{*2}(\gamma+1)}{(\gamma-1)^2F_0^3}\left[\left(F_1 + \frac{2c_j^{*2}}{\gamma-1}\right) - F_0F_2\right] \\ \phi^{(3)}(1) &= \frac{-2c_j^{*2}}{(\gamma+1)F_0^3}\left(F_1^2 - F_0F_2\right) \\ f^{(3)}(1) &= \frac{1}{2}\frac{(\gamma-1)\phi^{(3)}(1)}{\gamma} \end{aligned}\right\} \tag{9.8.23}$$

The zeroth-order boundary conditions (Eqs 9.8.20) are just the strong shock conditions. Note that the parameter c_j^{*2} appears in all the higher boundary conditions (i.e., Eqs 9.8.22–9.8.23) for the higher order solutions. For large values of the energy release E_0, $c_j^{*2} \to 0$ and all the higher-order boundary conditions become zero and the shock remains strong throughout its propagation. Eqs 9.8.22–9.8.23 also indicate that all the higher-order boundary conditions require the complete solution of the lower-order equations. For example, the first-order boundary conditions (Eq 9.8.22) contain F_0 which is determined from the zeroth-order energy integral (Eq 9.8.16). From Eq 9.8.14 we note that the zeroth-order shock decay coefficient $\theta^{(0)} = -\frac{1}{2}$ which gives the time exponent of the shock trajectory $N = 2/3$ (i.e., $x_s \sim At^N$). This is just the self-similar solution for strong planar blast waves. This is expected since for $x_s \to 0$, the shock wave is at the immediate vicinity of the wall at R_0 and curvature can be neglected and the strong blast is essentially planar.

After the zeroth-order solution has been determined, and F_0 evaluated from the energy integral, the first-order boundary conditions (Eq 9.8.21) can be found. The first-order equations (Eq 9.8.11) can then be integrated with a trial value of $\theta^{(1)}$. The correct solution is obtained when the value of $\theta^{(1)}$ chosen gives a solution that satisfies the boundary condition at the wall where the particle velocity must vanish. An alternate criterion for the iteration of the correct value of $\theta^{(1)}$ is when the trial value of $\theta^{(1)}$ agrees with the value of $\theta^{(1)}$ determined from the energy integral (i.e., $\theta^{(1)} = \frac{F_1}{2F_0}$) to within a prescribed accuracy. Second and higher-order solutions are found in a similar manner. In general, it is found that using the criterion of zero particle velocity as the wall is much more sensitive to iterate for the solution.

With the higher-order solutions determined and the F_n's evaluated, the shock trajectory can be obtained from the definition of the shock velocity, that is,

$$\dot{x}_s = \frac{\mathrm{d}x_s}{\mathrm{d}\tau},$$

or

$$\tau = \int_0^{x_s} \frac{\mathrm{d}x_s}{\dot{x}_s}.$$

Substituting the expansion for $\dot{x}_s^2$ in terms of x_s (i.e., Eq 9.8.13) into the above equation and integrating, one obtains

$$\begin{aligned}\tau &= -\int_0^{x_s} \mathrm{d}x_s \Big/ \left(\frac{F_0}{x_s} + F_1 + F_2 x_s + \cdots\right)^{\frac{1}{2}}, \\ &= B_0 |x_s|^{\frac{3}{2}} \left(1 + B_1 x_s + B_2 x_s^2 + B_3 x_s^3 + \cdots\right),\end{aligned} \tag{9.8.24}$$

where

$$B_0 = 2 \Big/ \left(\frac{1}{3F_0^{1/2}}\right),$$

$$B_1 = -\frac{3}{10}\frac{F_1}{F_0},$$

$$B_2 = \frac{3}{7}\left(\frac{3}{8}\frac{F_1^2}{F_0^2} - \frac{1}{2}\frac{F_2}{F_0}\right),$$

$$B_3 = -\frac{1}{3}\left(\frac{5}{16}\left(\frac{F_1}{F_0}\right)^3 - \frac{3}{4}\frac{F_1 F_2}{F_0^2} + \frac{1}{2}\frac{F_3}{F_0}\right).$$

For large energy release, $c_j^{*2} \simeq 0$ and the shock wave remains strong throughout. To illustrate the results for the perturbation solution for converging blast waves, the shock trajectories for the spherical case is shown in Fig. 9.8, for values of $C_j^{*2} = 0$,

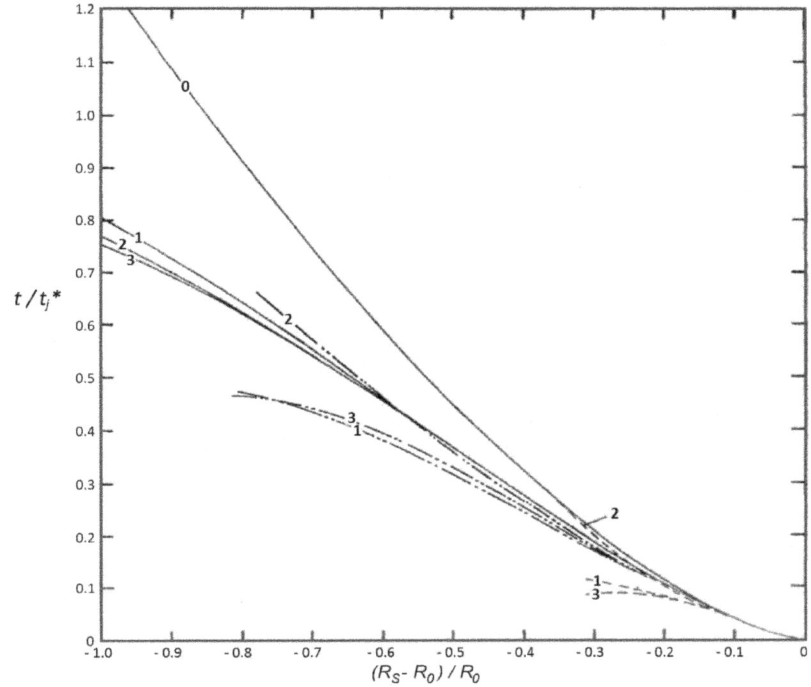

Figure 9.8 Shock Trajectories for Spherical Implosions at Different Initiation Energies. *Note:* Values of C_j^{*2}: ______0; ____...____, 0.1; __ __ __0.5. Zeroth-, first-, second-, and third-order solutions are shown

0.1, and 0.5. Also shown are the results for the different orders. For $C_j^{*2} = 0$, corresponding to a strong collapsing blast wave, we note that the third-order solution provides a fairly accurate description of the shock motion for the majority of the collapse. A change in the curvature of the shock trajectory can be observed indicating that the shock changes from an initially decaying wave to an accelerating wave as it converges. The zeroth-order solution corresponds to just a planar decaying blast wave. For finite values of C_j^{*2}, where the shock strength is finite, the range of validity of the third-order solution decreases and is only adequate close to the wall. A comparison of the shock trajectories for strong planar, cylindrical, and spherical (that is $C_j^{*2} = 0$) implosions is shown in Fig. 9.9. For the planar geometry, where there is no area convergence to amplify the shock, the blast decays like $R_S \sim t^{2/3}$ as given by the self-similar solution (Chapter 4). Note that to permit the comparison of the different geometries, the characteristic time t_j^* (a function of the geometry) must be referred to that of the planar case (i.e., t_0^*) in order to have the same initiation density per unit area for all the three geometries. From Eq 9.8.10, we observe that the same energy density $t_2^*/t_0^* = 1/\sqrt{3}$, $t_1^*/t_0^* = 1/\sqrt{2}$. Figure 9.9 further shows that all the trajectories coalesce in the vicinity of the wall as area convergence has not influenced the shock propagation as yet.

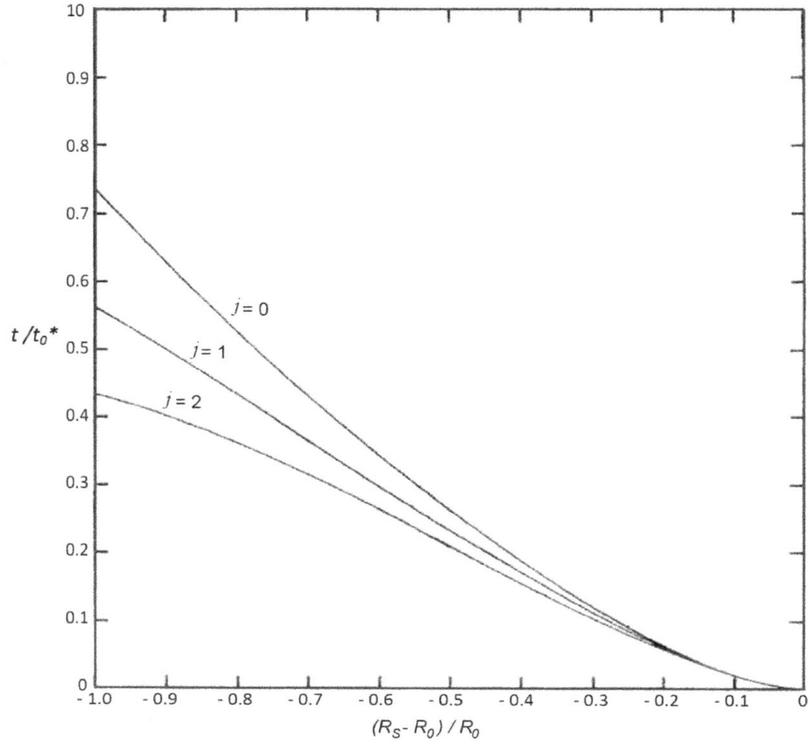

Figure 9.9 Comparison of Shock Trajectories for Strong Planar ($j = 0$), Cylindrical ($j = 1$), and Spherical ($j = 2$) Implosions ($C_j^{*2} = 0$)

The pressure profiles for strong ($C_j^{*2} = 0$) converging spherical blast waves are shown in Fig. 9.10. The zeroth order, corresponding to planar blast with no area convergence, is seen to compare well with the third-order solution. The initial pressure decay behind the blast is seen to indicate a decelerating blast wave. However, the pressure behind the shock increases as the shock propagates toward the center when the area convergence begins to cause the shock to accelerate.

The corresponding density profiles for strong converging spherical blast are shown in Fig. 9.11. The increase in density behind the shock as it amplifies when moving toward the center is more prominently illustrated than the pressure profile. For the final phases of the collapse near the center of symmetry, we redefine the variables as follows

$$\left.\begin{aligned} \psi(\xi, r_s) &= \frac{\rho(r,t)}{\rho_0} \\ \phi(\xi, r_s) &= \frac{u(r,t)}{\dot{R}_s} \\ f(\xi, r_s) &= \frac{p(r,t)}{\rho_0 \dot{R}_s^2} \\ \xi &= r/R_s \\ r_s &= R_s/R_0 \end{aligned}\right\} \tag{9.8.25}$$

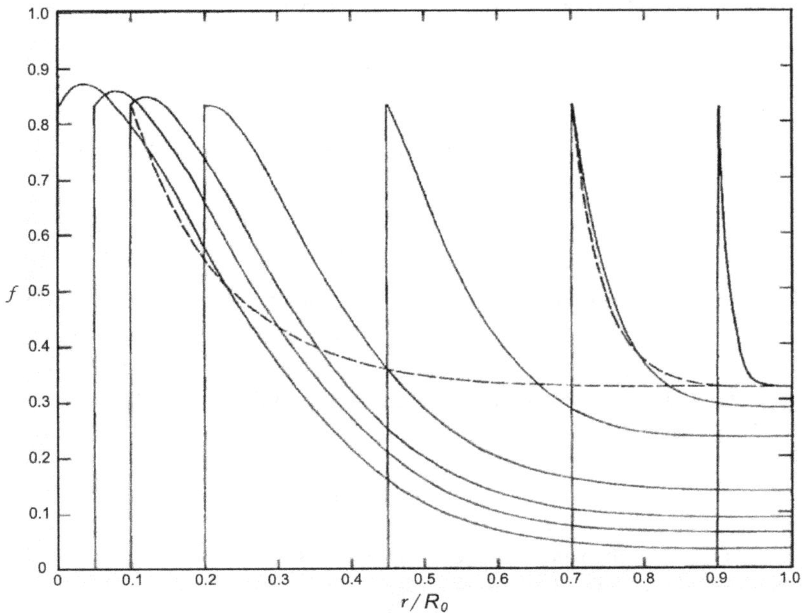

Figure 9.10 Normalized Pressure Profiles for Strong ($C_j^{*2} = 0$), Spherical Implosions, _ _ _, Zero Order; ____, Third Order

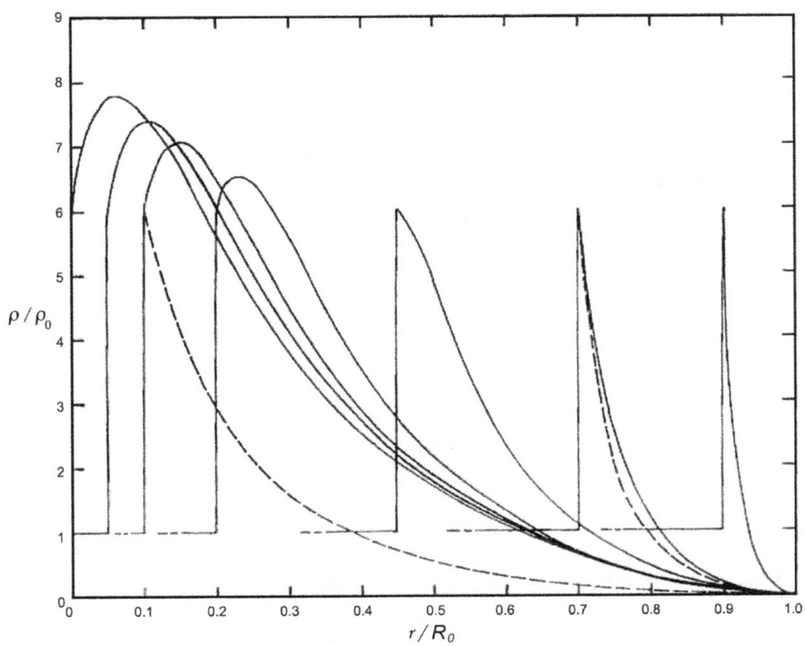

Figure 9.11 Normalized Density Profiles for Strong ($C_j^{*2} = 0$), Spherical Implosions, _ _ _, Zero Order; ____, Third Order

In terms of the above variables, the conservation equations can be written as

$$(\phi - \xi)\frac{\partial \psi}{\partial \xi} + \psi\frac{\partial \phi}{\partial \xi} + \frac{j\phi\psi}{\xi} = -r_s\frac{\partial \psi}{\partial r_s}, \tag{9.8.26}$$

$$(\phi - \xi)\frac{\partial \phi}{\partial \xi} + \frac{1}{\psi}\frac{\partial f}{\partial \xi} + \theta\phi = -r_s\frac{\partial \phi}{\partial r_s}, \tag{9.8.27}$$

$$(\phi - \xi)\frac{\partial f}{\partial \xi} + \gamma f\frac{\partial \phi}{\partial \xi} + \left(2\theta + \frac{\gamma j\phi}{\xi}\right) f = -r_s\frac{\partial f}{\partial r_s}, \tag{9.8.28}$$

where $\theta = \frac{r_s\ddot{r}_s}{\dot{r}_s^2}$. The flow is bounded by the shock front and the wall, thus $1 \leq \xi \leq \frac{R_0}{R_s}$. Initially at $t = 0$ where $R_s \simeq R_0$, $r_s \simeq 1$. Near the center of symmetry, $R_s \simeq 0$ and $r_s \simeq 0$. Thus $0 \leq r_s \leq 1$. For the final phase of the collapse, $r_s \to 0$ and $\theta \to \theta_0 =$ constant. The shock wave is strong and the boundary condition at the shock front is given by the limiting strong shock conditions, that is,

$$\psi(1) = \frac{\gamma + 1}{\gamma - 1}, \qquad \phi(1) = f(1) = \frac{2}{\gamma + 1}.$$

For $r_s \to 0$, the right hand terms in Eqs 9.8.26 and 9.8.27 drops out and solving for the derivatives with respect to ξ, we get

$$\phi' = \frac{\left(2\theta_0 f + \frac{\gamma j f\phi}{\xi}\right)\frac{1}{\psi} - \theta_0\phi(\phi - \xi)}{(\phi - \xi)^2 - \frac{\gamma f}{\psi}}, \tag{9.8.29}$$

$$f' = -\psi\left(\theta_0\phi + (\phi - \xi)\phi'\right), \tag{9.8.30}$$

$$\psi' = -\left(\psi\phi' + \frac{j\phi\psi}{\xi}\right)\Big/(\phi - \xi). \tag{9.8.31}$$

The above equations are identical to Eqs 9.8.26–9.8.28 given previously. The desired solution is determined by iterating for the value of $\theta^{(0)}$ that gives a regular solution at the singularity $(\phi - \xi)^2 - \frac{\gamma f}{\psi} = 0$.

The regularity condition to be satisfied is given by Eq 9.8.11, that is,

$$\theta_0 = \frac{\frac{\gamma j f^* \phi^*}{\xi^* \psi^*}}{\phi^*(\phi^* - \xi^*) - \frac{2 f^*}{\psi^*}},$$

where the asterisk $*$ is used to denote the values at the singularity $(\phi^* - \xi^*) - \frac{\gamma f^*}{\psi^*} = 0$.

The equation for the definition of θ, that is, $\theta = \frac{r_s\ddot{r}_s}{\dot{r}_s^2}$, can be integrated to yield the dependence of the shock strength $\dot{r}_s$ on the shock radius r_s. Doing so we obtain

$$\dot{r}_s = -(F_0)^{\frac{1}{2}}\, r_s^{\theta^{(0)}}, \tag{9.8.32}$$

where F_0 is a constant of integration. The negative sign is taken since the shock radius decreases with increasing time.

Integrating Eq 9.8.32 again gives the shock trajectory

$$r_s = A\left(1 - \frac{t}{t_c}\right)^N, \tag{9.8.33}$$

where

$$N = \frac{1}{1 - \theta^{(0)}},$$

$$A = \left[\frac{t_c (F_0)^{\frac{1}{2}}}{N} \right]^N .$$

In Eq 9.8.33, the constant t_c is chosen to yield $r_s = 0$, $t = t_c$. The constants $F^{(0)}$ and t_c cannot be found from the similarity solution and are determined by the initial conditions.

We may extend the validity of the self-similar solution to small, but finite, radius r_s, by seeking a perturbation solution in a small parameter $\varepsilon(r_s)$. From an examination of the basic equations and boundary conditions, we see that if the self-similar solution for $r_s \to 0$ is to be recovered as $\varepsilon \to 0$, then ε should be of the form

$$\varepsilon = r_s^{\alpha}, \tag{9.8.34}$$

where $\alpha = -2\theta^{(0)}$ is a positive number since $\theta^{(0)}$ is negative. We can write the following perturbation expressions for the variables

$$\dot{r}_s^2 = \frac{1}{\varepsilon} \sum_0^{\infty} F_n \varepsilon^n = \frac{1}{\varepsilon} (F_0 + F_1 \varepsilon + \cdots), \tag{9.8.35}$$

$$\theta(r_s) = \sum \theta_n \varepsilon^n = \theta_0 + \theta_1 \varepsilon + \cdots \tag{9.8.36}$$

$$\psi(\xi, r_s) = \Sigma \psi_n(\xi) \varepsilon^n = \psi_0(\xi) + \psi_1(\xi)\varepsilon + \cdots \tag{9.8.37}$$

$$\phi(\xi, r_s) = \Sigma \phi_n(\xi) \varepsilon^n = \phi_0(\xi) + \phi_1(\xi)\varepsilon + \cdots \tag{9.8.38}$$

$$f(\xi, r_s) = \Sigma f_n(\xi) \varepsilon^n = f_0(\xi) + f_1(\xi)\varepsilon + \cdots \tag{9.8.39}$$

From Eq 9.8.35 and the definition of θ, the coefficients θ_n in Eq 9.8.36 can be obtained as

$$\left. \begin{aligned} \theta_0 &= \frac{-\alpha}{2} \\ \theta_1 &= \frac{\alpha}{2} \frac{F_1}{F_0} \\ &\text{etc.} \end{aligned} \right\} \tag{9.8.40}$$

Substituting the perturbation expressions (i.e., Eqs 9.8.36–9.8.39) into the basic equations and boundary conditions, and collecting terms of the same order in ε, we obtain for zeroth order in ε, the self-similar equations given earlier by Eqs 9.8.29–9.8.31. For the first order in ε, we obtain

$$(\phi_0 - \xi)\psi_1' + \psi_0 \phi_1' = -\left[\psi_1(\phi_0' + \alpha) + \phi_1 \psi_0' + j(\phi_0 \psi_1 + \phi_1 \psi_0)/\xi \right], \tag{9.8.41}$$

$$(\phi_0 - \xi)\phi_1' + f_1'/\psi_0 = -\left[\phi_1(\phi_0' + \theta_0 + \alpha) + \theta_1 \phi_0 - \psi_1 f_0'/\psi_0^2) \right], \tag{9.8.42}$$

$$\begin{aligned} (\phi_0 - \xi)f_1' + \gamma f_0' \phi_1' = -\big[& f_1(\gamma \phi_0' + 2\theta_0 + \alpha) + 2\theta_1 f_0 + \phi_1 f_0' \\ & + \gamma j (f_0 \phi_1 + f_1 \phi_0)/\xi \big]. \end{aligned} \tag{9.8.43}$$

The first-order boundary conditions can be obtained by expanding the Rankine–Hugoniot equations (i.e., Eqs 9.8.6–9.8.8) using Eq 9.8.35 for $\dot{r}_s^2(\varepsilon)$. We may proceed to higher orders in ε in a similar manner.

The first-order solution is determined in an identical manner as the zeroth-order solution. From Eqs 9.8.41–9.8.43 we note that the singularity for the first-order equations is the same as the zeroth order (i.e., $(\phi - \xi)^2 - \frac{\gamma f_0}{\psi_0} = 0$). The corresponding regularity condition for the first order can be obtained from Eqs 9.8.41–9.8.43 as

$$\frac{1}{\psi_0}\left[\gamma f_1 \phi_0' + 2\theta_1 f_0 + \phi_1 f_0' + \gamma j(f_0\phi_1 + f_1\phi_0)/\xi\right]$$
$$- (\phi_0 - \xi)\left[\phi_1(\phi_0' - \theta_0) + \theta_1\phi_0 - \psi_1 f_0'/\psi_0^2\right] = 0.$$

The first-order solution is obtained by iterating for the value of θ_1 to satisfy the regularity condition at the singularity to within a prescribed accuracy.

We have obtained solutions for the initial and final phases of the implosion. A numerical solution is required in general to link the two solutions. However, the perturbation solution for the initial phase will provide an excellent starting condition for the numerical solution of the intermediate regime of the shock motion.

Bibliography

G.G. Bach and J.H.S. Lee. High order perturbation solutions for blast waves. *AIAA Journal*, 7(4):742–744, 1969.

G.G. Bach and J.H.S. Lee. An analytical solution for blast waves. *AIAA Journal*, 8(8):271–275, 1970.

G.G. Bach, K.W. Chiu, and J.H.S. Lee. Contribution to the propagation of non-ideal blast waves. I. Far field equivalency. 5th Int. Coll. Gasdynamics of Explosions and Reactive Systems, Bourges, France, 1975.

W.E. Baker. *Explosion in Air*. Wilfred Baker Engineering, San Antonio, 1974.

H. Bethe. Blast waves, chapter 4. approximation for small $\gamma - 1$. Technical report, Los Alamos Rept., 1947.

S.R. Brinkley and J.G. Kirkwood. Theory of the propagation of shock waves. *Physical Review*, 71(9):606–611, 1947.

D.S. Butler. Converging spherical and cylindrical shocks. Technical report, ARE Rept. 54/54, Kent, UK, 1954.

S. Chandrasekhar. On the decay of planar shock waves. Rept. 423, Aberdeen Proving Ground, 1943.

G. Chernyi. *Introduction to Hypersonic Flow*. Academic Press, New York, 1961.

W. Chester. The quasi-cylindrical shock tube. *Philosophical Magazine Series 7*, 45(371):1293–1301, 1954.

R.F. Chisnell. The motion of a shock wave in a channel, with applications to cylindrical and spherical shock waves. *Journal of Fluid Mechanics*, 2:286–298, 1957.

K.W. Chiu, J.H.S. Lee, and R. Knystautas. The blast waves from asymmetrical explosions. *Journal of Fluid Mechanics*, 82(1):193–208, 1977.

J.D. Cole. Newtonian flow theory for slender bodies. *Journal of the Aeronautical Sciences*, 24(6):448–455, 1957.

R.H. Cole. *Underwater Explosions*. Princeton University Press, Princeton, 1948.

R.H. Cole. *Underwater Explosions*. Princeton University Press, Princeton, 1961.

R. Courant and K.O. Friedrichs. *Supersonic Flow and Shock Waves*. Interscience, New York, 1948.

E.K. Dabora. Variable energy blast waves. *AIAA Journal*, 10(10):1384–1386, 1972.

M.N. Director and E.K. Dabora. Predictions of variable-energy blast waves. *AIAA Journal*, 15(9):1315–1321, 1977.

N.C. Freeman. An approach to the unsteady motion of a piston into a gas at rest with special reference to the pinch effect in electrical discharges. Aeronautical Research Council Report ARC-20,339 F.M. 2707, 1958.

N.C. Freeman. Newtonian theory of hypersonic flow at large distances from bluff axially symmetric bodies, in F.R. Riddel, editor, *Hypersonic Flow Research*, page 345, Elsevier, New York, 1962.

M.P. Friedman. An improved perturbation theory for shock waves propagating through nonuniform regions. *Journal of Fluid Mechanics*, 8:193–209, 1960.

K.O. Friedrichs. Formation and decay of shock waves. *Communications on Pure and Applied Mathematics*, 1(3):211–245, 1948.

I.I. Glass and J.P. Sislian. *Nonstationary Flows and Shock Waves*. Claredon Press, Oxford, 1994.

G.G. Guderley. Starke kugelige und zylindrische verdichtungstösse in der nähe des kugelmittelounktes bzw. der zylinderachse. *Luftfahrtforschung*, 19:302–312, 1942.

G.G. Guderley. Powerful spherical and cylindrical compression shocks in the neighborhood of the center of the sphere and of the cylinder axis. *Luftfahrtforschung*, 19:302–312, 1942.

Z. Han and X. Yin. *Shock Dynamics*. Kluwer Academic, 1993.

W.D. Hayes and F.R. Probstein. Hypersonic flow theory, Volume 1. *Inviscid Flows*. Academic Press, New York, 1966.

E. Jouguet. *Mécanique des explosifs*. Octave Doin et fils, Paris, 1917.

J.G. Kirkwood and S.R. Brinkley. Theory of propagation of shock waves from explosive sources in air and water. Technical Report A-318, NDRC Rept., 1945.

A.S. Kompaneets. A point explosion in an inhomogeneous atmosphere. *Soviet Physics Doklady*, 5:46, 1960.

V.I. Korobenikov and P.I. Clushkin. Thermal Prokladnol Mekhenika i Tekklinika Fizika, 4(48), 1963.

J. Kynch. Blast waves, in L. Howarth, editor, *Modern Developments in Fluid Dynamics: High Speed Flow*, volume 1, pp. 146–157, Oxford University Press, Oxford, 1953.

L.D. Landau. On shock waves at large distance from the place of their origin. *Journal of Physics-USSR*, 9:496–503, 1945.

L.D. Landau and E.M. Lifshitz. *Fluid Mechanics*. Pergamon Press, 1959.

D.D. Laumbach and R.F. Probstein. A point explosion in a cold exponential atmosphere. *Journal of Fluid Mechanics*, 35(01):53–75, 1969.

A. Lax. Decaying shock, a comparison of an approximate analytical solution with a finite difference method. *Communications on Pure and Applied Mathematics*, 1(3):247–257, 1948.

B.H.K. Lee. Non-uniform propagation of imploding shocks and detonations. *AIAA Journal*, 5(11):1997–2003, 1967.

J.H. Lee. The propagation of shocks and blast waves in a detonating gas. PhD thesis, McGill University, Montreal, Canada, 1965.

J.H. Lee. Gasdynamics of detonations. *Astronautica Acta*, 17:455–466, 1972.

J.H. Lee. *The Detonation Phenomenon*. Cambridge University Press, Cambridge, 2008.

J.H. Lee, B.H.K. Lee, and I. Shanfield. Two-dimensional unconfined gaseous detonation waves. *Symposium (International) on Combustion*, 10(1):805–815, 1965.

L. Lees and T. Kubota. Inviscid hypersonic flow over blunt-nosed slender bodies. *Journal of the Aeronautical Sciences*, 24(3):195–202, 1957.

C.K. Lewis. *Plane, Cylindrical, and Spherical Blast Waves Based Upon Oshima's Quasi-similarity Model*. Arnold Engineering Development Center, Tennessee, 1961.

M.J. Lighthill. The piston of the shock wave in certain aerodynamic problems. *Quarterly Journal of Mechanics and Applied Mathematics*, 1:309–318, 1948.

S.C. Lin. Cylindrical shock waves produced by instaneous energy release. *Journal of Applied Physics*, 25(1):54–57, 1954.

J.A. McFodden. Initial behavior of a spherical blast. *Journal of Applied Physics*, 23:269, 1952.

K. Oshima. Blast wave produced by exploding wires, in W. G. Chace and H. K. Moore, editors, *Exploding Wires*, volume 2, pages 159–174. Plenum Press, New York, 1962.

K. Oshima, K. Sugaya, M. Yamamoto, and T. Totoki. Diffraction of a plane shock wave around a corner. Rept. 393, Institute of space and Aero. Univ. Tokyo, 1965.

K. Oswatitsch. *Gasdynamics, English version by G. Kuerti*. Academic Press, New York, 1956.

W.J. Rae. Nonsimilar solutions for impact-generated shock propagation in solids final report. Technical Report NASA-CR-54251, Cornell Aeronautical Lab., 1965.

W.J. Rae and H.P. Kirchner. Final report on a study of meteoroid impact phenomena. Technical Report RM-1655-M-4, Cornell Aeronautical Lab., 1963.

M.H. Rogers. Similarity flows behind strong shock waves. *Journal of Mechanics and Applied Mathematics*, 11(4):411–422, 1958.

J. Rosciszewski. Calculations of the motion of non-uniform shock waves. *Journal of Fluid Mechanics*, 8:337–367, 1960.

M. Rosenbluth and M. Garwin. Infinite conductivity theory of the pinch, Technical report, Los Alamos Rept., LA-1850 1954.

A. Sakurai. On the propagation and structure of blast waves I. *Journal of the Physical Society of Japan*, 8(5):662–669, 1953.

A. Sakurai. On the propagation and structure of blast waves II. *Journal of the Physical Society of Japan*, 9(2):256–266, 1954.

A. Sakurai. On exact solution of the blast wave problem. *Journal of the Physical Society of Japan*, 10(9):827–828, 1955.

A. Sakurai. On the problem of a shock wave arriving at the edge of a gas. *Communications on Pure and Applied Mathematics*, 13:353–370, 1960.

A. Sakurai. Blast wave theory, in M. Holt, editor, *Basic Developments in Fluid Dynamics*, volume 1, pages 309–375. Academic Press, New York, 1965.

L.I. Sedov. *Similarity and Dimensional Methods in Mechanics*. Academic Press, New York, 1959.

A.H. Shapiro. *The Dynamics and Thermodynamics of Compressible Fluid Flow*. Ronald Press, New York, 1953.

R. J. Swigart. Third order blast wave theory and its application to hypersonic flow past blunt-nosed cylinders. *Journal of Fluid Mechanics*, 9:613–620, 1960.

G.I. Taylor. Air wave surrounding an expanding sphere. *Proceedings of the Royal Society of London. Series A, Mathematical and Physical Sciences*, 186(1006):273–292, 1946.

G.I. Taylor. The dynamics of the combustion of products behind plane and spherical detonation fronts in explosives. *Proceedings of the Royal Society of London. Series A. Mathematical and Physical Sciences*, 200(1061):235–247, 1950a.

G.I. Taylor. The formation of a blast wave by a very intense explosion. i. theoretical discussion. *Proceedings of the Royal Society of London. Series A, Mathematical and Physical Sciences*, 201:159–174, 1950b.

G.I. Taylor. The formation of a blast wave by a very intense explosion. ii. the atomic explosion of 1945. *Proceedings of the Royal Society of London. Series A, Mathematical and Physical Sciences*, 201:175–186, 1950c.

P.A. Thompson. *Compressible-Fluid Dynamics*. McGraw-Hill, New York, 1988.

S. von Hoerner. Lösungen der hydronamischen gleichungen mit linearem verlauf der geschwindigkeit. *Z. Naturforsch*, 10:687–692, 1955.

von Neumann. *The point source solution*, in *Collected Works of von Neumann*, Volume 6. Pergamon Press, Oxford, 1963.

R.L. Welsh. Technical report, University of California, Rept. No. AS-66.1, Berkeley, USA, 1966.

G.B. Whitham. On the propagation of weak shock waves. *Journal of Fluid Mechanics*, [illegible](3):290–318, 1956.

G.B. Whitham. On the propagation of shock waves through regions of non-uniform area or flow. *Journal of Fluid Mechanics*, 4:337–360, 1958.

G.B. Whitham. *Linear and Nonlinear Waves*. John Wiley & Sons, New York, 1974.

Ya.B. Zel'dovich. On the pressure and velocity distribution in detonative blast products, in particular for spherical propagation of the detonation wave. *ZhETF*, 12(9):389–406, 1942.

Y.B. Zel'dovich. Motion of a gas due to a pressure of short duration shock. *Soviet Physics-Acoustic (English Transl.)*, 2:25–35, 1956.

Y.B. Zel'dovich and Y.P. Raizer. *Physic of Shock Waves and High Temperature Hydrodynamic Phenomena*. Academic Press, New York, 1966.

Y.B. Zel'dovich and Y.P. Raizer. *Physics of Shock Waves and High-Temperature Hydrodynamic Phenomena*, volume 1. Academic Press, New York, 1967.

M.J. Zucrow and J.D. Hoffman. *Gasdynamics*, volume 1. John Wiley & Sons, New York, 1976.

Index

For EU product safety concerns, contact us at Calle de José Abascal, 56–1°,
28003 Madrid, Spain or eugpsr@cambridge.org.

www.ingramcontent.com/pod-product-compliance
Lightning Source LLC
LaVergne TN
LVHW062124200926
848484LV00007B/1756

* 9 7 8 1 1 0 7 1 0 6 3 0 7 *